Edited by
Genserik L.L. Reniers and
Luca Zamparini

Security Aspects of
Uni- and Multimodal
Hazmat Transportation Systems

Related Titles

Fries, R., Chowdhury, M., Brummond, J.

Transportation Infrastructure Security Utilizing Intelligent Transportation Systems

2009
ISBN: 978-0-470-28629-6

Bender, H.F., Eisenbarth, P.

Hazardous Chemicals
Control and Regulation in the European Market

2007
ISBN: 978-3-527-31541-3

Center for Chemical Process Safety (CCPS)

Guidelines for Chemical Transportation Safety, Security, and Risk Management

Second edition
2008
ISBN: 978-0-471-78242-1

Reniers, G.L.L.

Multi-Plant Safety and Security Management in the Chemical and Process Industries

2010
ISBN: 978-3-527-32551-1

Stoessel, F.

Thermal Safety of Chemical Processes
Risk Assessment and Process Design

2008
ISBN: 978-3-527-31712-7

Edited by Genserik L.L. Reniers and Luca Zamparini

Security Aspects of Uni- and Multimodal Hazmat Transportation Systems

WILEY-VCH Verlag GmbH & Co. KGaA

The Editors

Prof. Genserik L.L. Reniers
University of Antwerp
City Campus, Office B-434
Prinsstraat 13
2000 Antwerp
Belgium

Prof. Luca Zamparini
University of Salento
Dip. Studi Giuridici
Via per Monteroni snc
73100 Lecce
Italy

Library of Congress Card No.: applied for

British Library Cataloguing-in-Publication Data
A catalogue record for this book is available from the British Library.

Bibliographic information published by the Deutsche Nationalbibliothek
The Deutsche Nationalbibliothek lists this publication in the Deutsche Nationalbibliografie; detailed bibliographic data are available on the Internet at <http://dnb.d-nb.de>.

Print ISBN: 978-3-527-32990-8

Cover Design Formgeber, Eppelheim
Typesetting Toppan Best-set Premedia Limited, Hong Kong
Printing and Binding Markono Print Media Pte Ltd, Singapore

To the memory of my father, Giampaolo Zamparini, a hazmat transport logistician.

Luca Zamparini

Contents

Preface

In recent decades, transport security has gained increasing importance for both regulators and the industry, as well as for academia. The emphasis on this topic is partly due to specific security-related occurrences, but it is also linked to the ever-growing dimension of international transport and the consequent need to safeguard the most important hubs and links, given the worldwide consequences of a terrorist attack.

The transportation of chemicals and other hazardous materials is one industrial sector that is particularly prone to represent a terrorist target. The movement of such goods on roads, railways, waterways, and through pipelines, would thus need full protection that, however, appears to be both economically and practically infeasible.

Therefore, it is important to develop approaches, methods, and tools that allow informed and sound decisions to be made based on prioritized defensive and mitigating measures, and on the available resources.

Political and academic agendas are ever more filled with all possible aspects of transportation, and efforts are made to ensure that dangerous-materials transports are kept safe and secured. In order to reach the above-mentioned goals, different stakeholders with varying backgrounds and interests need to be involved in the prioritization and assessment process of hazmat transportation. This is because advancing security of hazmat transports is obviously very complex. A part of the complexity, but also of the opportunities, is provided by the available choices between different transport modes.

This book therefore investigates the latest economic and operational findings for the different modes, and discusses state-of-the-art insights and models for encompassing security within a multimodal framework and line of thinking. A discussion of the most relevant issues related to the movement of hazardous materials in the various transport modes and in a multimodal setting is proposed. Moreover, approaches from different regions around the world on how to deal with hazmat transport security and multimodality are given and elaborated.

In summary, trade-offs between security measures and industrial and civil liberties, and a multidisciplinary and multistakeholder approach are needed to truly advance hazmat transport security. While the book does not dictate any single model or method to be applicable for all transportation security problems, it does paint a complex picture of issues with possible solutions.

Antwerp and Lecce, November 11, 2011

Genserik Reniers
Luca Zamparini

List of Contributors

Marco Castro
ANT/OR – University of Antwerp
Operations Research Group
Prinsstraat 13
2000 Antwerp
Belgium

Wout Dullaert
Institute of Transport and Maritime
Management Antwerp
Keizerstraat 64
2000 Antwerp
Belgium
and
Antwerp Maritime Academy
Noordkasteel Oost 6
2030 Antwerp
Belgium

Fynnwin Prager
University of Southern California
School of Policy, Planning and
Development
University of Southern California
Los Angeles, CA 90089
USA

Juha Hintsa
Vrije Universiteit Brussel
Faculty of Economic Sciences
Department MOSI-Transport and
Logistics
Research group MOBI – Mobility and
Automotive Technology Bldg. M (231)
Pleinlaan 2
1050 Brussels
Belgium

Daniel Inloes, Jr.
University of Southern California
School of Policy, Planning and
Development
University of Southern California
Los Angeles, CA 90089
USA

Josip Kasum
University of Split
Zrinsko Frankopanska 38
Split 21000
Croatia

Amir Saman Kheirkhah
Bu Ali Sina University
Faculty of Engineering
Department of Industrial Engineering
Shahid Fahmideh Avenue
65174 Hamedan
Iran

Mark Lepofsky
Visual Risk Technologies, Inc.
1400 Key Blvd.
Suite 810
Arlington, VA 22209
USA

Cathy Macharis
Vrije Universiteit Brussel
Faculty of Economic Sciences
Department MOSI-Transport and Logistics
Research group MOBI – Mobility and Automotive Technology Bldg. M (231)
Pleinlaan 2
1050 Brussels
Belgium

Olivier Mairesse
Vrije Universiteit Brussel
Faculty of Economic Sciences
Department MOSI-Transport and Logistics
Research group MOBI – Mobility and Automotive Technology Bldg. M (231)
Pleinlaan 2
1050 Brussels
Belgium

Pablo Maya Duque
ANT/OR – University of Antwerp
Operations Research Group
Prinsstraat 13
2000 Antwerp
Belgium

Pamela Murray-Tuite
Virginia Tech University
Department of Civil and Environmental Engineering
7054 Haycock Road
Falls Church, VA 22043
USA

Henk Neddermeijer
Kernel Group
Postbus 357
2740 AJ Waddinxveen
The Netherlands

Paola Papa
University of Salento
Dipartimento di Studi Giuridici
Via per Monteroni, snc
73100 Lecce
Italy

Paul W. Parfomak
Specialist in Energy and Infrastructure Policy
Congressional Research Service
101 Independence Avenue SE
Washington, DC 20540
USA

Genserik Reniers
University of Antwerp
Antwerp Research Group on Safety and Security (ARGoSS)
Prinsstraat 13
2000 Antwerp
Belgium
and
HUB, KULeuven
Centre for Economics and Corporate Sustainability (CEDON)
Stormstraat 2
1000 Brussels
Belgium

Mohja Rhoads
University of Southern California
School of Policy, Planning and Development
312 RGL
University of Southern California
Los Angeles, CA 90089
USA

Lisa Schweitzer
University of Southern California
School of Policy, Planning and Development
312 RGL
Los Angeles, CA 90014
USA

Kenneth Sörensen
ANT/OR – University of Antwerp
Operations Research Group
Prinsstraat 13
2000 Antwerp
Belgium

Joseph S. Szyliowicz
University of Denver
Josef Korbel School of International Studies
Ben M. Cherrington Hall
201 South Gaylord Street
Denver, CO 80208
USA

Christine Vanovermeire
ANT/OR – University of Antwerp
Operations Research Group
Prinsstraat 13
2000 Antwerp
Belgium

Koen Van Raemdonck
Vrije Universiteit Brussel
Faculty of Economic Sciences
Department MOSI-Transport and Logistics
Research group MOBI – Mobility and Automotive Technology Bldg. M (231)
Pleinlaan 2
1050 Brussels
Belgium

Manish Verma
Memorial University
Faculty of Business Administration
BN. 3017
St. John's, NL
Canada A1B 3X5

Bert Vernimmen
Institute of Transport and Maritime Management Antwerp
Keizerstraat 64
2000 Antwerp
Belgium

Vedat Verter
McGill University
Desautels Faculty of Management
1001 Sherbrooke St., West Montreal,
Quebec, H3A 1G5 Canada

Pero Vidan
Faculty of Maritime Studies
University of Split
Zrinsko Frankopanska 38
Split 21000
Croatia

Luca Zamparini
Università del Salento
Dipartimento di Studi Giuridici
Via per Monteroni snc
73100 Lecce
Italy
and
Faculty of Social, Political and Regional Studies
Cittadella della Ricerca di Brindisi-Mesagne
72100 Brindisi
Italy

Part One
Introductory Section

1
Editorial Introduction

Luca Zamparini and Genserik Reniers

The last decade has witnessed an increasing worldwide concern for security-related issues. Terrorist attacks have hit several regions of the world (the United States, Great Britain and Spain in Europe, Indonesia and so on) and this has raised the interest towards security in a dramatic way at all levels of government (local, national, and international). In this context, it is important to consider that it has been computed that, between 1970 and 2010, about 6% of all terrorist attacks have targeted transport means and infrastructures[1]. It then becomes relevant that the improved security in transport represents one of the key topics in the agendas of counterterrorist agencies worldwide. In the United States, for example, one of the first actions undertaken soon after the 9/11 attacks was the creation of the Transportation Security Administration as an agency of the US Department of Homeland Security.

At the European Union level, the strategy to augment the security of transportation has mainly been based on a series of regulations, directives, and proposals aimed at enhancing the security levels of the various transport modes and infrastructures. On the Asian continent, international policies related to security have mostly taken the form of agreements on themes as transit of goods and people and on more general security cooperation issues.

This increased political concern on transportation security has been somewhat matched by researchers in various disciplines (among them, economics, law, engineering, political science) that have tried to analyze the current security systems and devices and the transport networks at the local, national, and international scales to propose viable alternatives to strengthen the security procedures, without hampering in a marked way the need for seamless and efficient transport flows.

One of the basic issues that part of the literature has been trying to clarify is the conceptual heterogeneity between safety and security in transportation. The first term refers to the absence (or, more properly, the minimization) of all dangers,

1) See the Global Terrorism Database at http://www.start.umd.edu/gtd/features/GTD-Data-Rivers.aspx (accessed 11 July 2011).

Security Aspects of Uni- and Multimodal Hazmat Transportation Systems, First Edition. Edited by Genserik L.L. Reniers, Luca Zamparini.

risks, injuries, and fatalities that may depend on accidental and unintended events related to inadvertent or hazardous behavior. On the other hand, security can be defined as the prevention of and protection against deliberate actions that aim at generating (mass) fatalities, disruption of services, and economic and social distress. From the scientific viewpoint, the main difference between safety and security is represented by the fact that the former can be analyzed by means of statistical and probabilistic techniques, while the latter (given its very low frequency and the fact that terrorist acts are intentional) cannot be treated with the same tools and requires, for example, the use of cost–benefit analyses in order to estimate the economic and social incentives to raise the level of security, and, for example, of game theory to mimic the strategic interactions between people and organizations planning and executing terrorist actions and counterterrorist agencies.

A segment of the transport sector that is particularly prone to the generation of mass fatalities in the case of a terrorist attack and that requires particular strategies, actions, and protocols in order to guarantee its degree of security is definitely represented by the hazardous materials (so-called "hazmat") business. The large quantities of explosives and of chemical, radioactive, and poisonous goods that are shipped every day within and among countries represent both a necessity for the productive sector and a concern for security agencies and personnel. In this context, the efficiency and the security of transport do not seem to represent a trade-off but rather a conjoint necessity and goal. Particular care has to be paid to the planning, implementation, and monitoring of hazmat transport activities both at the company and at the Government level. Routes have to be selected with particular care in order to minimize the possible effects of terrorist acts given that shipments of hazardous materials can represent both targets and weapons of mass destruction. Moreover, the security of the transport nodes (ports, warehouses, logistic platforms, and so on) is probably even more important than the monitoring of transport flow (Lewis, 2008). It is thus relevant to consider the hazmat transportation not just as a series or set of unimodal activities but rather as an integrated multimodal system.

By taking into consideration all of the above-mentioned topics, the present book, whose structure is sketched in Figure 1.1, aims at covering both the unimodal and the multimodal issues related to hazmat transportation. The first introductory section will provide a description of the history and importance of hazmat transportation and of the main economic themes and models that have been proposed in the literature in order to analyze this sector. The second section, in line with a part of the traditional literature on hazmat transportation, will analyze the various transport modes that are concerned with hazmat transportation (road haulage, railways, inland waterways, and pipelines) from a unimodal perspective. The third section will offer a multimodal perspective both in terms of formal models and of empirical evidence. The fourth section will present a series of country case studies (Italy, The Netherlands, the United States, and Iran) in order to ascertain the similarities and homogeneities in several geographic regions around the world, subject to different economic and social contexts.

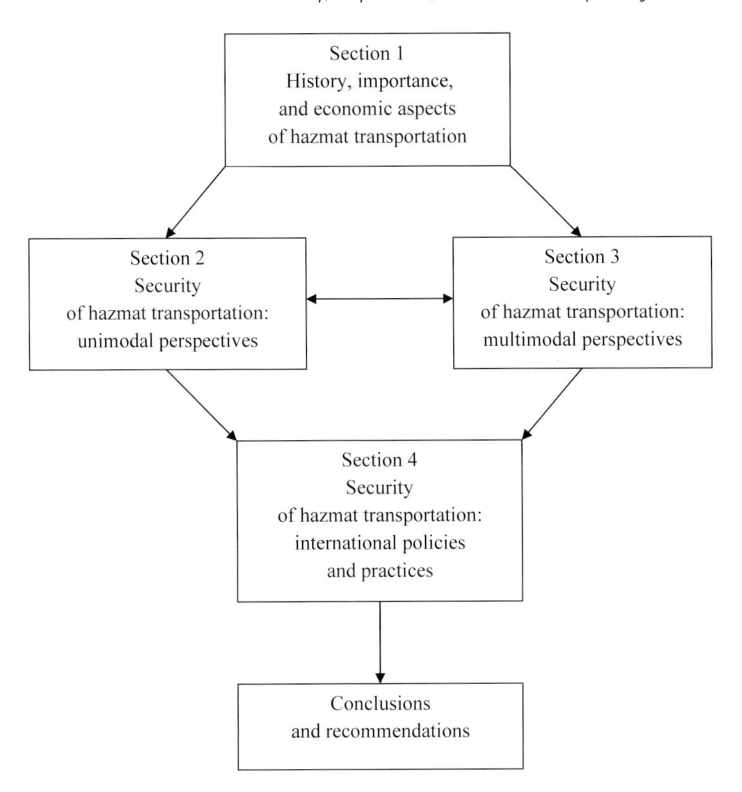

Figure 1.1 Structure of the book.

The next paragraphs will offer an outlook of the fourteen contributions that constitute the various sections of the book. In order to provide an introduction to the various chapters, a description of the main themes that are dealt with in each one will be given. Moreover, the main topics that constitute the book (security, efficiency, unimodal and multimodal approaches) will be highlighted.

1.1 History, Importance, and Economic Aspects of Hazmat Transportation

This first, introductory, section is constituted by two contributions. The first one, *History and Importance of Hazmat Transportation*, clarifies what are the categories of hazardous materials and their paramount importance for the competitiveness and development of both the industrial and the service sector. In this respect, the operational risks that characterize hazmat transportation and the implied difficulties in their planning are discussed jointly with the need for collaboration among security experts and transport specialists. The chapter also sketches a chronology of the main hazmat transport accidents since the late 1970s and the role played by security issues, especially after 9/11, for a renewed interest in hazmat-transport-related research and indicates sustainability as the other key issue. After proposing the latest security-related events, the chapter concludes by mentioning the economic relevance of some terrorist attacks, the benefit in terms of advancement in technology that can spur from security-related research.

The second contribution of this section, *Economic Issues in Hazmat Transportation*, provides a brief description of the relevance of the hazmat transport market both in the US and in the European Union in order to highlight the relevance of the hazmat-transport market, where the diffusion of multimodal transport is still limited. Moreover, it surveys recent models that have been proposed in order to analyze the risk assessment, the routing/scheduling, and the allocation problem in the case of hazardous materials where heterogeneous analytical settings are discussed and the scarcity of models considering security is underscored.

1.2
Security of Hazmat Transportation: Unimodal Perspectives

The second section of the book is composed of four chapters. The first one, *Security of Hazmat Transports by Road*, pinpoints the large heterogeneity that exists in terms of road haulage of hazardous materials. It also describes the various possible truck types that can be used for hazmat. It then analyzes the various responsibilities that pertain to motor carriers in terms of general security issues, risk assessment, attack profiles, and training of personnel. The chapter also lists and considers all possible threats that may be related to unauthorized access (in loading docks, storage facilities, vehicles, and so on) and all possible profiles of *en route* security. Lastly, it describes the technologies that can be usefully employed to counter intended unlawful acts with a particular emphasis on the Hazmat Transport Vehicle Tracking System introduced in Singapore in 2005.

The second chapter of this section, *Security Aspects of Hazmat Transport Using Railroad*, discusses the reasons why security is very relevant in the case of rail transport, given the degree of interconnectivity of its arcs and nodes and the large number of entry points for perspective terrorists. Coherently with the previous chapter, it then proposes all possible sources of risk. Moreover, it stresses the importance in identifying the critical points in the railway system, the route risks, and the probabilities and consequences of attacks. The chapter then comments on the adopted steps in risk-management strategies (information sharing and coordination, policing and surveillance and routing of hazmat) and on the necessity to implement further steps (interdiction models, tank car design, and placement of hazmat railcars).

The third chapter of this section, *Security of Hazmat Transports by Inland Waterways*, introduces the most important legislation that is related to this segment of the transport market and describes the causes that can lead to a lack of safety of a vessel in navigation, emphasizing the security-related ones (terrorism, vandalism, pilferage). It also proposes the current regulations and practices enacted in order to increase security. Moreover, it proposes the strategy to further improve inland waterways transportation security suggested by the "International Ship and Port Facility Security Code in Inland Waterways" and its three levels of operations (vessels, organizations, and ports) and other connected initiatives (as container monitoring).

The last contribution of this section, *Security of Hazmat Transports by Pipeline,* proposes an outlook of the hazmat pipeline infrastructure around the world, specifying the regional shares and the transported materials. It then describes the security risks to hazmat pipelines and, especially, the commodity thefts and the global terrorist attacks and incidents in the last decade. Moreover, it estimates the costs and impacts of pipeline security incidents and the range of measures that have been implemented by pipeline operators and government agencies. The second part of the chapter describes the US strategy and security programs in the last decades and emphasizes the need for international cooperation and exchange of available sensitive information.

1.3
Security of Hazmat Transportation: Multimodal Perspectives

The third section of the book is constituted by four chapters. The first chapter, *Multimodal Transport: Historical Evolution and Logistics Framework,* constitutes a general introduction to the multimodal perspectives discussed in this section. It describes the trends in multimodal transport in the European Union, in the United States, and in the ASEAN (Association of South-East Asian Nations) region, providing the economic rationale for its diffusion and the likely future trends in the next decades. The chapter also proposes a logistics model that provides a standard framework and the important variables that have to be taken into account in the choice among several multimodal transport alternatives.

The second contribution of this section, *Multimodal Analysis Framework for Hazmat Transports and Security,* proposes a review of the literature related to hazmat transport emphasizing the role of multicriteria analysis and of multicriteria routing models to analyze this market. It then proposes a model for the calculation of multimodal hazmat-transport risk and considering the probability of occurrence of a catastrophic incident as the result of combination of a general probability and of a locality parameter. It also considers the impact of these events on the basis of the involved transport mode. The chapter then shifts its attention to intended incidents due to terrorists or activists, and proposes a model that considers security as one of the most relevant parameters for modal choice.

The following chapter of this section, *Metaheuristics for the Multimodal Optimization of Hazmat Transports,* introduces the metaheuristics technique and clarifies the rationale for its use in the case of multimodal hazmat transportation optimization problems. In this context, it discusses the role of multilevelness and multiobjectivity, and surveys the contributions that have used metaheuristics for multimodal transportation in general and for hazmat transportation in particular. Lastly, it proposes a peculiar metaheuristic for hazmat transportation in the case of an intermodal network.

The last chapter of this section, *Freight Security and Livability: US Toxic and Hazardous Events from 2000 to 2010*, is based on the previous contributions and discusses the implications of transport consolidation and distribution strategies on the local communities that live around important hazmat-transportation hubs. It also discusses the interactions and relationships among land use, infrastructure location, and industrial organization. The chapter then tests these assumptions on the basis of the events that have occurred in the last decade in California and relates them to the location of multimodal hub facilities. It then compares the evacuation, the environmental damage, the time loss, and the total damage on the basis of the transport mode and of the hazardous materials class in terms of response, property and remediation costs.

1.4 Security of Hazmat Transportation: International Policies and Practices

The fourth section of the book is based on four contributions. The first one, *Security of Hazmat Transport in Italy*, ascertains the economic significance of hazmat transport in Italy with a set of statistics related to the last decade. It also describes the Italian legal framework on hazmat security for the various transport modes and provides the list of the most relevant and recent cases where the security measures have been effective. The second chapter, *Security of Hazmat Transport in the Netherlands from a Security Practitioner's Point of View*, discusses the peculiarities of The Netherland's stance on security and the role played by its infrastructure as a possible risk factor. It then analyzes the security issues that pertain to transport and logistics in this country, and discusses the network of private and public organizations that are involved in the degree of security in the country. The third contribution, *Safeguarding Hazmat Shipments in the US: Policies and Challenges*, compares the pre-9/11 and the post-9/11 situations and policies, and provides detailed statistics of the magnitude of the hazmat-transport business in the US. It then examines the vulnerabilities of the rail sector and the envisaged policies, protocols and emergency planning and response. It further analyses the role and the interactions among the federal government, the local governments and the private sector, and the situation and issues related to road haulage. The last contribution of this section, *Security of Hazmat Transports in Iran*, provides a description of the most relevant hazmat-transport-related accidents in the last decade, and proposes and discusses a list of all the strengths and weaknesses related to this sector and the optimal policies to implement in order to increase the security level.

Bibliography

Lewis, T. G. (2008) *Critical Infrastructure Protection in Homeland Security. Defending a Networked Nation*, John Wiley & Sons, Inc., Hoboken, NJ.

2
History and Importance of Hazmat Transportation

Genserik Reniers

2.1
Introduction

Transporting chemical substances to serve the chemicals using industries (such as e.g., petrochemical plants, pharmaceutical companies but also industries manufacturing, for example, paints, varnishes, soaps, detergents, etc.) is needed for the storage, production, and distribution of raw materials, base chemicals, intermediates, etc., within and across regional, national and international borders. Daily transportation activities of such so-called "hazardous materials" (*hazmat*) or "dangerous goods" via roads, railways, inland waterways and pipeline networks are essential to national economies, and such transportation is even crucial to – and a necessary condition for – a healthy world economy and our modern-day lives. In fact, service industries, including financial, medical and social services, are only made possible by the wealth-producing activities of production industries and the transportation activities between the numerous chemical plants, storage and production centers, etc. The competitiveness of all these sectors is partly dependent on the efficient supply of chemical products. It is for that reason that the chemical industry, including hazmat transportation, has been described as the "anchor" of a modern economy (Howitt, 2000; McKinnon, 2004; Schreckenbach and Becker, 2006).

In general, the chemical industry has a high intensity in terms of its generation of transportation and logistics activities. On average, the European chemical industry spends about 8 to 10 per cent of its total turnover on logistics and supply chain activities. This was estimated at 60 billion euro and represents 1.5 billion tons of movement per year in 2005 (Braithwaite, 2005). In 2008, 82 billion tonne-km of dangerous goods were transported in EU-27 (ES1, 2009). Hauliers in the 6 main European economies, Germany, Spain, Italy, France, the UK and Poland accounted for nearly 75% of all these transports of dangerous goods in Europe. It is obvious that hazmat transportation is crucial for healthy local and international economies.

In the current global context of industrial activities, the prosperity and the wealth of nations may indeed be heavily affected if a major disruption of the hazmat

Security Aspects of Uni- and Multimodal Hazmat Transportation Systems, First Edition. Edited by Genserik L.L. Reniers, Luca Zamparini.

transportation system would occur to some extent. The growing amounts of dangerous goods produced and transported indicate that the dependence of the global economy on the hazmat-transportation system is substantially increasing over time. Hazmats include explosives, flammables, corrosive, poisonous or infectious substances, and toxic waste among others. Hence, as a natural byproduct, different types of operational risks associated with hazmat transports are also becoming ever more important and ever more difficult to assess and to manage. Private companies, governments and regulatory organizations worldwide are discussing safety and security issues concerning hazmat and debating the extent of its inherent operational risks in today's paradigm of sustainable development.

Millions of hazardous-materials shipments move daily through worldwide transportation systems. Operational risks associated with these shipments can be generated by either accidental circumstances, or by intentional actions. As already discussed in the introductory Section, the former risks are termed *safety risks*, while the latter are known as *security risks* (CCPS, Center for Chemical Process Safety, 2003). Safety and security are thus two related concepts (Aven, 2006, 2008; Holtrop and Kretz, 2008; Johnston, 2004) but they have a different basis (Holtrop and Kretz, 2008): safety is nonintentional, whereas security is intentional (Aven, 2006, 2008; Holtrop and Kretz, 2008; Hessami, 2004; EWS (Early Warning System): available online via http://vbo-feb.be/search_results/title-EWS/). This implies that in the case of security an aggressor is present (Johnston, 2004; Randall, 2008; George, 2008) who is influenced by the physical environment and by personal factors (Randall, 2008). The hazmat-transportation system obviously goes hand in hand with all kinds of safety and security risks and problems such as potential pollution, overuse of road and railway infrastructure, etc.

Compared with the average nonhazmat truck shipments accident rate (which is 0.73 per million vehicle miles in the USA), hazardous materials truck shipment accidents remain rare events (on average 0.32 per million vehicle miles in the USA) (Federal Motor Carrier Safety Administration, 2001). Moreover, although the number of deaths and injuries due to all traffic accidents dwarfs the fatalities and injuries figures due to hazmat-related accidents, public concern about the risk of hazmat incidents is rather intense. Even minor incidents involving chemical freights strongly attract the attention of the general public, policy makers and industrialists. This is primarily due to the involuntary nature of the risk and the potential for significant consequences in the case of such accidents. Despite the fact that there is genuine public concern about the operational risks associated with hazmat-transportation systems, complete protection of these systems is economically and practically infeasible. Therefore, there is a need for a multidisciplinary perspective on hazmat security, thereby taking the different transport modes into account, and based on a systems approach. Safety and security management specialists need to collaborate with transport (and other) experts, using for example, the mathematical techniques of game-theory, operations research, etc. Such a holistic managerial approach should lead to more objective and cost-effective assessments and prioritizations of defenses and countermeasures against existing hazmat-transportation threats.

From an operations research viewpoint for example, Erkut and Verter (1998) indicate that risks associated with the transport of hazardous materials make the planning of such transports very difficult. The authors suggest that different risk models usually select different "optimal" paths for a hazmat shipment between a given origin–destination pair. Furthermore, the optimal path for one model could perform very poorly under another model. Researchers and practitioners must thus pay considerable attention to the modeling of safety and security risks in hazardous-materials transportation.

2.2 History of Hazmat-Transportation Research

The attention to academic and industrial research regarding risks associated with the hazmat-transportation system and the carrying out of safety studies on dangerous freight traffic, dates back to the early 1980s. Some important transport accidents from each of the transport modes leading to the increased attention at that time, were, among others, a tank car accident in Los Alfaques (Spain) in 1978 claiming 216 fatalities, a waterway accident in Bantry Bay (Ireland) in 1979 including 50 deaths, a railroad accident in 1981 in San Luis Potosi (Mexico) claiming some 20 lives, and a pipeline accident in Cubatos (Brazil) in 1984 that took 89 lives. In the late 1990s, there was a slight slow-down in attention due to difficulties of gathering accurate and relevant information and data. Recently, the topic of hazmat-transportation research has regained interest, among others due to the increased importance of the factors *security* (since 9/11) and *sustainability*. Centrone, Pesenti, and Ukovich (2008) indicate that the number of peer-reviewed papers in international journals published between 1982 and 2007 peaked in mid-1990s, decreased till 2006, and has grown again since 2007. Security and sustainability are the driving forces for this renewed attention on the topic, as will be explained below.

The first (and foremost) driving force for increased research attention towards hazmat transportation is *security*: public transport security has become a focus of public concern and academic research in the Western World since the attacks of 9 September 2001 (New York WTC terrorist attack), 11 March 2004 (Madrid railway terrorist attack) and 7 July 2005 (London subway terrorist attack). Various aspects of public transport security have been investigated and analyzed by a number of studies.

After the 9/11 attacks in New York City, it was quickly recognized that, besides public means of transport, hazmat vehicles could also be desirable targets for terrorists. For example, public concerns in the United States led to requirements such as hazmat truckers submitted to fingerprinting and criminal background checks (Glaze, 2003). Recently, new threats stimulated interest in hazmat-transport security as a standalone academic research issue, and some studies have been carried out on the subject. Hazmat-transportation security problems on highways received the most attention in operations research literature (Centrone, Pesenti, and Ukovich, 2008). In contrast, security issues related to hazmat transports through a pipeline network, as well as multimodal transports of dangerous goods, have received little attention thus far. However, the hazmat-transportation system is so hard to secure and its security aspects are so difficult to investigate, not so much because tackling methods of securing one of the transportation modes is difficult, but because tackling all of the modes' security at the same time, due to their interconnectivity with one another, is a complex challenge.

The second factor enhancing researchers' attention towards hazmat transportation is *sustainability* and the worldwide climate change debate. Potential impacts of climate change on transportation are geographically widespread, modally diverse, and may affect both transportation infrastructure and operations. The global economy of the twenty-first century will need the natural resources of the Arctic and subarctic. Growing evidence of climate change indicates that ice-free navigation seasons will probably be extended and thinner sea ice will probably reduce constraints on winter ship transits. Furthermore, reduced ice cover would permit better access to polar regions and longer shipping seasons for, among others, locating, extracting and transporting hazardous materials. Reduced sea ice is likely to allow increased offshore extraction of oil and gas (ACIA [Arctic Climate Impact Assessment], 2004). Besides environmental-protection issues and pollution problems, issues of safety and security will arise. An increase in marine access for transport and offshore development will give rise to an increase in potential conflicts among competing users of Arctic waterways and coastal seas. Terrorism and piracy may be expected in the Arctic region as well.

At present, each year, close to 2 billion tons of hazardous materials is produced in the United States alone. The amount of hazardous-materials shipments that are shipped each year in the USA is approximately 3 billion tons, since each shipment is moved several times before reaching its destination. Over the period 1997–2002, the total cargo had risen 20% in the United States (USA Census bureau, 2002. Available via: http://www.census.gov). The total cargo in Europe in the same period had risen some 14% (TREN [Transport and Energy], 2005). Approximately 6% of the total cargo represents dangerous goods transports. Hazmat transports are thus essential for the world economy and their importance is continuously rising. Research as regards the security of the hazmat-transportation system from a multimodal viewpoint therefore deserves much more attention from private and public research institutes.

2.3
Importance of Research on Hazmat Transportation and Associated Risks

As indicated in the previous section, global trade has resulted in more worldwide national and transborder shipments of raw materials, dangerous goods, hazardous wastes, etc., than ever before. The volume of traffic and the speed with which the transports move continue to increase in both developed and developing countries. Although approximately 67 000 individuals worldwide were either killed or injured by terrorist attacks in 2007 alone (National Counterterrorism Center, 2008), remarkably, there have been relatively few mass-casualty disasters associated with hazardous-materials transports in the Western world, especially when one considers the volumes carried per container, the number of shipments each year, and the direct proximity of the public. Historically, worldwide, only a very limited number of intentional incidents causing mass casualties have happened, when expressed per billion ton-miles.

A recent well-known security-related accident concerns a passenger train that was derailed by a powerful home-made bomb explosion on 28 November 2009 in Russia, thereby killing some 30 people and injuring some 100 people. Another security-related incident concerns an attempt to derail a passenger train in July 2010. In this sabotage incident, the railway was cut over 1.5 meters between Lelystad and Almere, two cities in The Netherlands. A well-known example of a security-related pipeline disaster happened in the Abule Egba district of Lagos, Nigeria. More than 260 people were killed and over 60 were injured after a gasoline pipeline exploded. The pipeline was ruptured deliberately and had been tapped by thieves for months prior to the accident. In another recent security-related pipeline accident on 11 August 2010, two people were killed and one wounded after an oil pipeline exploded in southeast Turkey in an apparent terrorist attack by the PKK.

Although until the present the worldwide number of security-related accidents involving hazmat transports is rather limited (if Iraq and Afghanistan are not taken into account), a wide variety of security threats do exist against hazmat transports. A chemical shipment can fall victim to a number of different types of attacks inspired by terrorism. A country's air, land and marine transport systems are designed for accessibility and efficiency, two characteristics that make them highly vulnerable to terrorist attacks. Such possible attacks, combined with recent (safety-related) accidents, are further changing the perception and changing the tolerance of managing the risk of hazardous-material transportation operations. The vulnerabilities of the transportation infrastructure and the potential consequences of hazmat transportation incidents put pressure on politicians and industrialists, and urge governmental policy makers to take action, resulting in worldwide calls for legislation directed at increased security, rerouting of hazardous-material shipments with emphasis on security, and the push for application of inherently safer technologies (leading to enhanced security as a natural byproduct).

Nonetheless, until the present, the terrorist threat to the hazmat-transportation system has gained far less attention from political and industrial policy makers than the passenger-transport one, since very few terrorist organizations have made a serious attempt to either target hazmat transports, or to use hazmat transports as their means of attack in countries belonging to the Western world.

Terrorist actions may be targeted at obtaining as many human fatalities as possible, or at causing as much economic devastation as possible or they may be aimed at realizing both these objectives. It is important to note that under the present circumstances, the hazmat-transportation system, in the USA as well as in Europe or any other part of the world, if under attack in an intelligent way, can be employed by terrorists to attain both these goals simultaneously.

Particularly in the transportation system, escalation effects from an attack may be very far-reaching. As an example, in the Madrid terrorist attack not only were 191 people killed and some 1800 injured, but the drop in the European markets following the Madrid bombings was calculated to be around $55 billion. As another example, the estimated cost on the entire supply chain of a weapon of mass destruction shipped via container is some $1 trillion (Zamparini, 2010).

Therefore, to proactively limit the potential number of human fatalities and the huge economic costs, it is crucial that industrial and academic research is carried out from a multidisciplinary, multimodal and systemic viewpoint, to identify, to assess and to prioritize transborder transportation-system security risks, and to take proactive actions and countermeasures accordingly. Counteracting the vulnerability of the chemical supply chain and finding the best possible ways to make it more resilient and secure is of utmost importance.

2.4 Conclusions

As the amounts of dangerous goods transported continues to rise in both developed and developing countries, and as population characteristics continue to increase throughout the world, and the distance between the public and the hazardous-materials transportation operations continuously diminishes, controlling and managing risks within the multimodal hazardous-materials transportation system becomes ever more important for practitioners as well as for policy makers. In particular, risks associated with intentional acts are the subject of the emerging research field devoted to the hazmat-transportation system.

The threat from terrorism is gaining importance. Terrorism strikes without warning, at any given location, at any given time. Passenger transports remain attainable targets of choice, while it is very likely that hazmat transports are the subject of future attacks. From a multidisciplinary researchers' perspective as well as from a policy makers' viewpoint, the *transportation system security* research area therefore needs to be treated as a mature and extremely important domain with a direct impact on the global economic market.

Bibliography

ACIA (Arctic Climate Impact Assessment) (2004) *Impacts of A Warming Arctic: Arctic Climate Impact Assessment*, Cambridge University Press, Cambridge, United Kingdom.

Aven, T. (2006) A unified framework for risk and vulnerability analysis covering both safety and security. *Reliability Engineering and System Safety*, **92** (6), 745–754.

Aven, T. (2008) Identification of safety and security critical systems and activities. *Reliability Engineering and System Safety*, **94** (2), 404–411.

Braithwaite, A. (2005) *Maximising performance: The power of supply chain collaboration*, The European Petrochemical Association (EPCA), Brussels, Belgium.

CCPS, Center for Chemical Process Safety (2003) *Guidelines for Analyzing and Managing the Security Vulnerabilities of Fixed Chemical Sites*, American Institute of Chemical Engineers, New York, USA.

Centrone, G., Pesenti, R., and Ukovich, W. (2008) Hazardous materials transportation: a literature review and an annotated bibliography, in *Advanced Technologies and Methodologies for Risk Management in the Global Transport of Dangerous Goods* (eds C. Bersani, A. Boulmakoul, E. Garbolino, and R. Sacile), The NATO Science for Peace and Security Programme, IOS Press, Amsterdam, The Netherlands, pp. 33–62.

Erkut, E. and Verter, V. (1998) Modeling of transport risk for hazardous materials. *Operations Research*, **46** (5), 625–642.

ES1 (2009) Global economic crisis hits European road freight transport in the fourth quarter of 2008. Accessed via http://epp.eurostat.ec.europa.eu/cache/ITY_OFFPUB/KS-SF-09-086/EN/KS-SF-09-086-EN.PDF (accessed April 2011).

Federal Motor Carrier Safety Administration (2001) *Comparative Risks of Hazardous Materials and Non-Hazardous Materials Truck Shipment Accidents/Incidents*, Battelle, Ohio, USA.

George, R. (2008) Critical infrastructure protection. *International Journal for Critical Infrastructure Protection*, **1**, 4–5.

Glaze, M. (2003) New security requirements for hazmat transportation. *Occupational Health and Safety*, **72** (9), 182–185.

Hessami, A.G. (2004) A system framework for safety and security: the holistic paradigm. *System Engineering*, (7). doi: 10.1002/sys.10060

Holtrop, D. and Kretz, D. (2008) *Onderzoek security & safety: een inventarisatie van beleid, wet- en regelgeving*. Nederland: Arcadis.

Howitt, S. (2000) *Keynote Report: The Chemical Industry*, Chartered Institute of Marketing.

Johnston, R.G. (2004) Adversarial safety analysis: borrowing the methods of security vulnerability assessments. *Journal of Safety Research*, **35** (3), 245–248.

McKinnon, A. (2004) *Supply chain excellence in the European chemical industry*, The European Petrochemical Associatiuon (EPCA), Brussels, Belgium.

National Counterterrorism Center (2008) 2007 Report on Terrorism, USA.

Randall, L.A. (2008) *21st Century Security and CPTED*, CRC Press, Boca Raton, USA.

Schreckenbach, T. and Becker, W. (2006) Chemicals – Driving innovation in other industries, in *Value Creation. Strategies for the Chemical Industry*, 2nd edn (eds F. Budde, U.-H. Felcht, and H. Frankemölle), Wiley-VCH Verlag GmbH, Weinheim, Germany, pp. 41–52.

TREN (Transport and Energy) (2005) *Evaluation of EU Policy on the transport of Dangerous Goods since 1994*, TREN/E3/43-2003 (Final Report), Pira International.

Zamparini, L. (2010) Transport Security in EU and US: Competing or Complementary visions? Nectar workshop on transport security, Lecce, Italy, 5–6 Feb 2010.

3
Economic Issues in Hazmat Transportation

Luca Zamparini

3.1
Introduction

The previous chapter has emphasized the importance of hazmat transport given its flows and relevance in all manufacturing processes and the renewed interest for security. The present chapter is intended as a complement to the previous one and it has a two-fold aim. On the one hand, it will provide detailed statistics related to two important geographic areas (the United States and the European Union) in order to ascertain the diffusion of hazardous-material transport and the relative importance of the various hazmat classes. In this context, the partition among the different transport modes and the multimodal option will also be considered. On the other hand, it will offer an overview of models that have taken into account hazardous-material transport. According to previous literature reviews, the models will be partitioned according to risk assessment, routing/scheduling and allocation problems. The aim of this section is to highlight the main theoretical models that have been proposed, the transport modes considered and the related main findings.

The structure of the chapter is as follows. Section 3.2 is based on a series of datasets that elicit the importance and relevance of hazmat transport in the United States and in the European Union. Section 3.3 offers a survey of recent hazardous-materials transport models. The last section concludes the chapter.

3.2
Hazmat Transportation in the United States and in the European Union

The present section aims at introducing and commenting on recent data related to hazmat transportation in the European Union and in the United States, two geographic areas for which it was possible to retrieve systematic data on this peculiar market.

Analyzing the data for US in 2007 (Table 3.1), it is first possible to notice that the large majority of hazmat shipments is accomplished through single modes

Security Aspects of Uni- and Multimodal Hazmat Transportation Systems, First Edition. Edited by Genserik L.L. Reniers, Luca Zamparini.

Table 3.1 Hazmat shipments in US by transportation mode (2007).

	Value		Tons		Ton miles		Miles
Transportation mode	**$ Billion**	**%**	**Millions**	**%**	**Billions**	**%**	**Avg dist. per shipment**
All modes, total	1448	100	2231	100	323	100	96
Single modes, total	1371	94.6	2112	94.6	279	86.3	65
Truck	837	57.8	1203	53.9	104	32.2	59
Rail	69	4.8	130	5.8	92	28.5	578
Water	69	4.8	150	6.7	37	11.5	383
Air	2	0.1	na	na	na	na	1095
Pipeline	393	27.2	629	28.2	na	na	na
Multiple modes, total	71	4.9	111	5.0	43	13.3	834
Unknown and other modes	7	0.5	8	0.4	1	0.5	58

Source: US Department of Transportation 2010.

(94.6% of the total). Among the various transport modes, the majority of shipments require road haulage and the second most used mode is pipeline. Rail and internal waterways have equal shares in terms of transported value and air transport only accounts for a residual share in terms of value.[1] Multimodal transport of hazmat goods equaled around 5% in terms of value and of quantity, while a larger share (13.3%) is related to ton-miles quantity, given that the average distance per shipment is 834 km when multiple modes are considered, while for single modes it is only 65 km. Similar considerations can be applied within modes, where the share of rail transport in ton-km (28.5%) is relatively close to that of road haulage (32.2%).

When hazard classes are considered (Table 3.2), it emerges that the large majority of shipments in this market is related to flammable liquids (class 3) that accrue to roughly 80% in terms of both value and quantity. The second most important class is then represented by gases followed by corrosive materials (class 8) and by miscellaneous dangerous goods (class 9). All other classes have residual shares.

From the data provided by the US Department of Transportation (Table 3.3), it is possible to elicit the evolution of hazmat transport safety between 1980 and 2009. The four punctual observations do not allow a determined pattern to be highlighted. However, both the number of total incidents and of the highway ones seem comparable at the beginning and at the end of the period. Given the increase in the amount of transported goods, this implies an increased relative safety of hazmat transports.

When the European Union is considered, it emerges that the total amount of dangerous goods transported by road in 2007 accounted for about 4.1% of the total

1) Data related to quantities were not available for this transport mode.

Table 3.2 Hazmat shipments in US by hazard class (2007).

Hazard class	Description	Value		Tons		Ton miles	
		$ Billions	%	Millions	%	Billions	%
Class 1	Explosives	12	0.8	3	0.1	<1	<0.1
Class 2	Gases	32	9.1	251	11.2	55	17.1
Class 3	Flammable liquids	1170	80.8	1753	78.6	182	56.1
Class 4	Flammable solids	4	0.3	20	0.9	6	1.7
Class 5	Oxidizers and organic peroxides	7	0.5	15	0.7	7	2.2
Class 6	Toxic (poison)	21	1.5	11	0.5	6	1.8
Class 7	Radioactive materials	21	1.4	<1	<0.1	<1	<0.1
Class 8	Corrosive materials	51	3.6	114	5.1	44	13.7
Class 9	Miscellaneous dangerous goods	30	2.1	63	2.8	23	7.1
Total		1448	100	2231	100	323	100

Source: US Department of Transportation 2010.

Table 3.3 Hazmat transportation incidents in US (1980–2009).

	1980	1990	2000	2009
Total	15 719	8879	17 557	14 777
Air	223	297	1 419	1 357
Highway	14 161	7296	15 063	12 691
Rail	1 271	1279	1 058	641
Water	34	7	17	88
Other	30	0	0	na

Source: US Department of Transportation 2010.

road haulage in the EU27 (excluding Italy and Malta). Very similarly to the situation in the United States, the largest share in ton-km is represented by flammable liquids (58%) followed by gases (12%), corrosives (10%), and miscellaneous dangerous substances (7%). The other hazmat classes account for minor shares (see Figure 3.1).

In the case of rail transport in the European Union (Figure 3.2), hazmat represented about 14% of the total freight transport by rail in the EU27 (excluding Bulgaria, Cyprus, Malta and Romania) in 2006; a remarkably larger share than the one related to road haulage.

In terms of relative importance of the various hazmat classes, the composition for rail transport is homogeneous to the one related to road haulage with

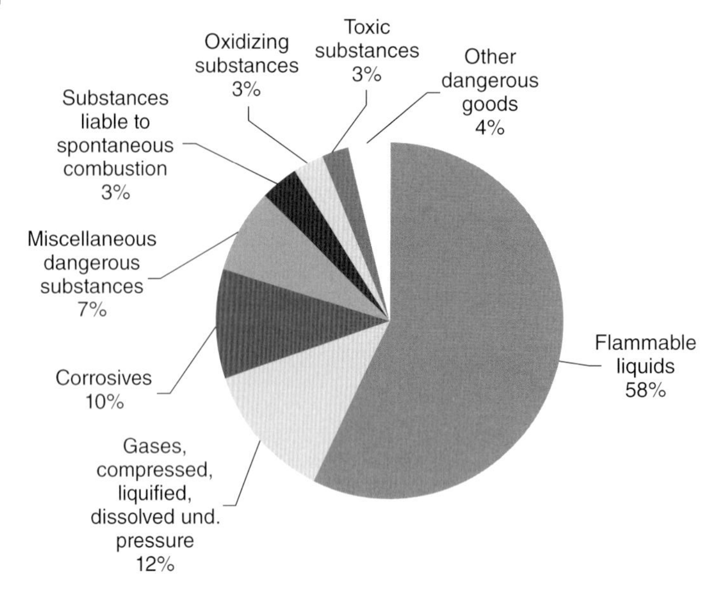

Figure 3.1 Dangerous goods transported by road, EU27*, 2007 (% tkm). Source: Eurostat, 2009.

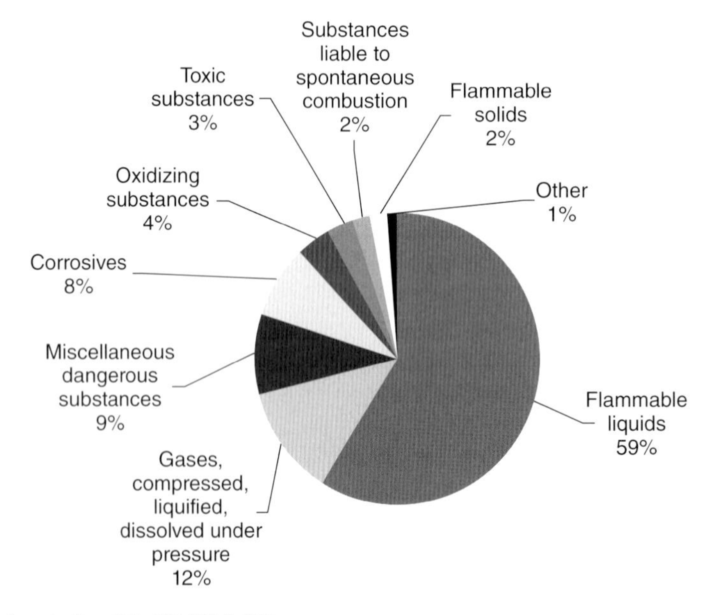

Figure 3.2 Dangerous goods transported by rail, EU27*, 2006 (% tkm). Source: Eurostat 2009.

flammable liquids, gases, and miscellaneous dangerous substances accounting for 80% of the overall ton-km transported in this region.

It is also possible to consider the spatial distribution of hazmat rail transport in the various European countries (Table 3.4). It is possible to notice that the two most relevant markets are represented by Germany and Poland, followed by the Czech Republic and Romania and by the three Baltic countries. These latter countries are characterized by the highest degree of hazmat specialization (88% in Estonia, 80% in Latvia and 66% in Lithuania). No spatial pattern seems to emerge for the countries that have the lowest rail-shipped quantities (Luxembourg, Greece, Norway, Denmark and Ireland). On the other hand, Sweden, Norway and Denmark are the three countries with the lowest specialization indexes.

Given the relevance of the hazmat transport phenomenon, the scientific literature has attempted, in the last decades, to provide models that try to optimize

Table 3.4 Hazmat rail transport in European countries (2006).

Country	Total tons of hazmat cargo (1000 tons)	Share of hazmat of total ton of cargo
Germany	117.68	34
Poland	82.89	53
Estonia	53.93	88
Czech Republic	45.82	47
Romania	40.46	63
Latvia	38.98	80
Lithuania	33.15	66
France	19.50	18
Austria	14.65	17
Hungary	13.57	29
Slovakia	12.06	23
Netherlands	9.20	30
Belgium	8.71	14
Bulgaria	8.47	40
Finland	6.53	15
Croatia	4.46	29
Turkey	4.15	21
Spain	3.99	16
Italy	3.53	5
Slovenia	3.41	20
Sweden	2.60	4
Portugal	2.25	23
Luxembourg	0.97	8
Greece	0.50	13
Norway	0.48	2
Denmark	0.15	2
Ireland	0.00	–

Source: author's elaboration based on Kolbenstevdt and Amundsen 2009.

several issues of hazmat transportation. The next section will provide a review of the main lines of research and of the related models.

3.3
Models of Hazmat Transport

Two literature surveys that were conducted at the beginning of the 1990s (List *et al.*, 1991) and of the last decade (Luedtke, 2002), emphasized that the main focuses of research for models of hazmat transport are risk analysis, routing/scheduling, and facility location. The former line of research deals with risk assessment by considering the likelihood of accidents and their related outcomes in terms of damages and of fatalities (see, among others, Rhyne, 1994; Erkut and Verter, 1995; Bonvicini, Leonelli, and Spadoni, 1998). The second class of models consider the choice of the appropriate route by taking into account a variety of objectives that may include costs, safety and security issues (see, e.g., Patel and Horowitz, 1994; Nozick, List, and Turnquist, 1997; Frank, Thill, and Batta, 2000). The latter models estimate the optimal location of productive and or warehouse that may ship or receive hazardous materials (see, e.g., Seton, Fitterer, and Harris, 1992; Ben-Moshe, Katz, and Segal, 2000). Other models have combined the last two issues, routing/scheduling and facility location, in order to provide comprehensive models that would dead with both topics (see, among others, Current and Ratick, 1995; Giannikos, 1998; Min, Jayaraman, and Srivastava, 1998).

The following subsections will discuss some of the most recent contributions along the above-mentioned research lines, trying to underscore the analytical techniques used, the most relevant themes and the main findings.

3.3.1
Risk Assessment in Hazmat Transport

A paper by Verter and Kara (2001) introduced the Geographical Information System (GIS) in order to implement on a large scale the risk assessment of shipments in a multiple origin–destination setting. The interesting issue is that in this paper the optimization process takes into account a traditional transport economics measure (minimization of the distance) jointly with a series of hazmat-transport-peculiar issues (population exposure, number of people to be evacuated, and probability of an incident). The model was applied to the Canadian provinces of Ontario and Quebec and took into account truck shipments.

Another paper that took into account road haulage of dangerous goods was proposed by Fabiano *et al.* (2005). It aimed at presenting a model that would not only consider risk assessment but also the optimization of emergency planning by stating the role of infrastructural, meteorological, and traffic factors. On the basis of a dataset collected in Northern Italy, they proposed a model based on graph theory in order to minimize the level of risk under some acceptability parameters. Along the main research lines, is a paper by Brown and Dunn (2007) that, on the basis of a quantitative risk assessment approach, proposes a method to evaluate to what distance an emergency response should be planned in order to protect people from an inhalation hazard that follows the accidental release of toxic materials. The database considered is related to the various states that compose the US and the transport modes considered are road haulage and rail transport.

A further paper considering the latter transport mode (rail) is that by Gheorghe *et al.* (2005) that is based on Swiss data. The proposed approach to risk assessment is based on a deductive loss of containment frequency calculation. It then implements a series of analytical models in order to assess the various types of impact of an incident (i.e. fatalities and acute intoxication) and, lastly, by recurring to further analytical techniques, it tries to identify the spots in a territory with the highest likelihood of accidents. The above-mentioned techniques are then integrated in order to provide a systematic risk assessment.

The paper by Jo and Ahn (2005) takes into account pipeline transport and considers a model that is intended to reduce the level of risk as a key variable that should be taken into account in the planning and building stages of a new pipeline. It stresses the importance of considering not only individual but also societal risk due to cumulative fatal length and failure rate. The dataset used in the paper is based on data at the European level.

3.3.2
Routing/Scheduling and Hazmat Transport

Contrary to the previous approach, all papers related to routing/scheduling models of hazmat transport that will be surveyed in this subsection are related to road haulage. In this context, a paper by Huang, Long, and Liew (2003) has taken into account security considerations in proposing a model for hazmat routing in the case of road haulage. The analytical model is based on the GIS and on the Analytic Hierarchy Process that allow determination of the weights of the various possible criteria (and among them security expressed as the risk of hijack) in order to compute the total costs connected with the various alternative routes. The area study is a neighborhood of the Singapore city area.

A different approach has been adopted in the paper by Zografos and Androutsopoulos (2004) that uses an heuristic algorithm to solve a problem of distribution routes, characterized as a vehicle-routing problem with time windows, that have to minimize a cost and an individual risk function, when risk is computed as the probability of death due to hazmat accidents. On the same line of research, a paper by Meng, Lee, and Cheu (2005) considers data from Singapore and proposes a multiobjective model that deals with transportation networks characterized by multiple time-varying attributes (travel time, population exposed and so on). Lastly, a contribution by Erkut and Alp (2007) considers jointly a routing and scheduling problem where not only population exposure but also accident rates and links durations vary with the time of day. They also test their dynamic programming algorithm with a case study based on data related to an Italian region (Lazio).

A different approach is taken by Bell (2006) who considers the case of a risk-averse planning of routing of hazardous materials and suggests that the way to enhance safety is to opt for sharing shipments among routes. The analytical model is based on the short-path algorithm.

3.3.3 Allocation Problem in Hazmat Transport

Fewer contributions have been related to the allocation problem in the case of hazmat transport. A recent paper by Jiahong and Bin (2010) has considered the problem of locating emergency response centers in order to maximize their speediness to get to possible accident sites. The solving algorithm was based on the multiobjective linear programming method that was tested with an hypothetical example that starts with the consideration of three different criteria (cost, time and coverage) and moves then to the above-mentioned multiobjective in order to show that the latter is more comprehensive.

3.4 Concluding Remarks

The previous sections have provided a brief description of the relevance of the hazmat-transport market both in the US and in the European Union and a survey of recent models that have been proposed in order to analyze the risk assessment, the routing/scheduling and the allocation problem in the case of hazardous materials.

It emerges clearly from Section 3.2 that the hazmat transport is a very relevant market where the diffusion of multimodal transport is still below its potential and further investments and efforts in order to increase the market share of this segment are needed. Moreover, at the European Union level, the diffusion of rail transport is not evenly distributed, with a concentration on the Germany–Poland–Czech Republic area and on the Baltic countries.

The models surveyed in Section 3.3 have provided very different analytical settings and economic contexts where they have been tested. However, only one model considered security explicitly. However, several models take into account emergency-response procedures that may be usefully employed not only for safety-related accidents but also in case of (breach in) security episodes. The next chapters of the book are aimed at providing a panorama of the relevance of security in both unimodal and multimodal transportation.

Bibliography

Bell, M.G.H. (2006) Mixed routes strategies for the risk-averse shipment of hazardous materials. *Networks and Spatial Economics*, **6**, 253–265.

Ben-Moshe, B., Katz, M.J., and Segal, M. (2000) Obnoxious facility location: complete service with minimal harm. *International Journal of Computational Geometry and Applications*, **10** (6), 581–592.

Bonvicini, S., Leonelli, P., and Spadoni, G. (1998) Risk analysis of hazardous materials transportation: evaluating uncertainty by means of fuzzy logic. *Journal of Hazardous Materials*, **62** (1), 59–74.

Brown, D.F. and Dunn, W.E. (2007) Application of a quantitative risk assessment method to emergency response planning. *Computers and Operations Research*, **34**, 1243–1265.

Current, J. and Ratick, S. (1995) A model to assess risk, equity and efficiency in facility location and transportation of hazardous materials. *Location Science*, **3** (3), 187–201.

Erkut, E. and Alp, O. (2007) Integrated routing and scheduling of hazmat trucks with stops en route. *Transportation Science*, **41** (1), 107–122.

Erkut, E. and Verter, V. (1995) A framework for hazardous materials transport risk assessment. *Risk Analysis*, **15** (5), 589–601.

Eurostat (2009) *Panorama of Transport*, 2009 edn, Eurostat, Luxembourg.

Fabiano, B., Currò, F., Reverberi, A.P., and Pastorino, R. (2005) Dangerous goods transportation by road: from risk analysis to emergency planning. *Journal of Loss Prevention in the Process Industries*, **18**, 403–413.

Frank, W.C., Thill, J.C., and Batta, R. (2000) Spatial decision support system for hazardous materials truck routing. *Transportation Research Part C: Emerging Technologies*, **8** (1–6), 337–359.

Gheorghe, A.V., Birchmeier, J., Vamanu, D., Papazoglou, I., and Kröger, W. (2005) Comprehensive risk assessment for rail transportation of dangerous goods: a validated platform for decision support. *Reliability Engineering & System Safety*, **88** (2), 247–272.

Giannikos, I. (1998) A multiobjective programming model for locating treatment sites and routing hazardous wastes. *European Journal of Operational Research*, **104** (2), 333–342.

Huang, B., Long, C.R., and Liew, Y.S. (2003) GIS-AHP model for hazmat routing with security considerations. Paper presented at the IEEE 6th International Conference on Intelligent Transportation Systems, Shanghai, China.

Jiahong, Z. and Bin, S. (2010) A new multi-objective model of location-allocation in emergency response network design for hazardous materials transportation. Paper presented at the 2010 IEEE International Conference on Emergency Management and Management Sciences, pp. 246–249, Beijing, China.

Jo, Y.-D. and Ahn, B.J. (2005) A method of quantitative risk assessment for transmission pipeline carrying natural gas. *Journal of Hazardous Materials*, **A123**, 1–12.

Kolbenstevdt, M. and Amundsen, A. (2009) *Rail Freight Security Practices. Literature Review and Background for a Retrack Security Survey*, Working Package 7 of the Retrack (Reorganization of Transport Networks by Advanced Rail Freight Concepts) EU Sixth Framework Program Project.

List, G.F., Mirchandani, P.B., Turnquist, M., and Zografos, K. (1991) Modeling and analysis for hazardous materials transportation: risk analysis, routing/scheduling and facility location. *Transportation Science*, **25**, 100–114.

Luedtke, J. (2002) *Hazmat Transportation and Security: Survey and Directions for Future Research*, Department of Industrial and System Engineering, Georgia Institute of Technology.

Meng, Q., Lee, D.-H., and Cheu, R.L. (2005) Multiobjective vehicle routing and scheduling problem with time window constraints in hazardous material transportation. *Journal of Transportation Engineering*, **131** (9), 699–707.

Min, H., Jayaraman, V., and Srivastava, R. (1998) Combined location-routing problems: a synthesis and future research directions. *European Journal of Operational Research*, **108** (1), 1–15.

Nozick, L.K., List, G.F., and Turnquist, M.A. (1997) Integrated routing and scheduling in hazardous materials transportation. *Transportation Science*, **31** (3), 200–215.

Patel, M.H. and Horowitz, A.J. (1994) Optimal routing of hazardous materials considering risk of spill. *Transportation Research Part A: Policy and Practice*, **28** (2), 119–132.

Rhyne, W.R. (1994) *Hazardous Materials Transportation Risk Analysis: Quantitative Approaches for Truck and Train*, Van Nostrand Reinhold, New York.

Seton, W., Fitterer, R.S., and Harris, T. (1992) Step up to the challenge: hazmat storage. *Chemical Engineering*, **99** (6), 118.

US Department of Transportation (2010) *Freight Facts and Figures 2010*, US DOT, Washington.

Verter, V. and Kara, B.H. (2001) A GIS-based framework for hazardous materials transport risk assessment. *Risk Analysis*, **21** (6), 1109–1120.

Zografos, K.G. and Androutsopoulos, K.N. (2004) A heuristic algorithm for solving hazardous materials distribution problems. *European Journal of Operational Research*, **152**, 507–519.

Part Two
Security of Hazmat Transports: Unimodal Perspectives

4
Security of Hazmat Transports by Road

Mark Lepofsky

4.1
Introduction

As with all modes of transportation, shipments of hazardous materials by road are vulnerable to various security threats and pose significant security challenges to the stakeholders in the transportation supply chain. Hazardous-materials transports by road differ from those by other modes primarily in that they provide the greatest level of access to terrorists or those wishing to use the materials for hostile or criminal purposes. Rail, water, and pipeline transportation, for example, constrain shipments to defined corridors; whereas road networks can take hazardous materials to almost any location. Trucks, in general, are not viewed suspiciously in most areas and can be used to deliver hazardous materials directly to many desirable targets.

Another unique aspect of the nature of road transport versus other modes that carry hazardous materials is its sheer diversity. In the United States alone, there are approximately 77 thousand hazardous materials carriers, both for-hire and private (those that transport only company-owned commercial materials)[1] Road transport is used to carry hazardous materials more than any other mode in the United States, measured by tons or ton-miles (Bureau of Transportation Statistics, 2010). Presumably, this holds true for most other countries as well, but data are not readily available. In the European Union, for example, 77.5 per cent of all freight was carried by road, but this statistic is not reported for dangerous goods[2]. For example, Latvia and Lithuania do show greater percentages of overall freight transported by rail than by road.

1) Federal Motor Carrier Safety Administration, Analysis and Information Online, Hazardous Materials (HM) Overview, http://ai.fmcsa.dot.gov/HazmatStat/Default.aspx (accessed 26 February 2011) United States Department of Transportation.
2) European Union, Eurostat. http://epp.eurostat.ec.europa.eu/portal/page/portal/transport/data/database, Modal Split of Freight Transport table (accessed 18 April 2011).

Security Aspects of Uni- and Multimodal Hazmat Transportation Systems, First Edition. Edited by Genserik L.L. Reniers, Luca Zamparini.

Hazardous materials of all classifications are transported by road and in a wide variety of packagings and vehicle configurations. However, not all materials transported are very attractive to a terrorist; security efforts should be focused on the materials and shipments that have both the greatest attractiveness to a terrorist and that can cause the greatest potential consequences. Some basic level of security, of course, is appropriate for all hazardous materials – and for all motor-carrier shipments in general. The typical approach used today is to employ a threat- or risk-based approach to identifying and prioritizing appropriate measures to mitigate the security risk of road transport of hazardous materials.

In the United States, there has been a strong history of safety initiatives focused on the hazardous materials community, including transportation. One prominent example is the chemical industry's Responsible Care® initiative and its Distribution Code of Management Practices[3]. Many of the safety practices are relevant and useful for security. However, often safety and security can be competing, such as when facilitating the sharing of information about the hazardous materials being transported in a vehicle with emergency responders to improve safety may also provide that same valuable information to terrorists, from whom it should be concealed. Many of these safety initiatives have evolved to include security, such as the Responsible Care® Security Code. In Europe, industry has also been active in promoting improved security of hazardous-materials transportation. The European Commission's Industry Guidelines for the Security of the Transport of Dangerous Goods by Road had the involvement of many organizations, including AISE (International Association for Soaps, Detergents and Maintenance Products), CEFIC (European Chemical Industry Council), CEPE (European Council of the Paint, Printing Ink and Artists' Colours Industry), CLECAT (European Association for Forwarding, Transport, Logistics and Customs Services), ECTA (European Chemical Transport Association), EFMA (European Fertilizer Manufacturers Association), FECC (European Association of Chemical Distributors), FIATA (International Federation of Freight Forwarders' Associations), and IRU (International Road Transport Union).

4.2 Hazmat Truck Types

Hazardous materials are commonly carried in many different types of vehicles. Some of the trucks and trailers most commonly used throughout the world are shown in Table 4.1. Of course, there are many types of smaller "straight" trucks of various lengths and configurations that may carry smaller quantities of hazardous materials.

3) The Codes of Management Practice have evolved into the current Responsible Care® Management System, initiated in 2002 (American Chemistry Council, http://www.americanchemistry.com, accessed 28 February 2011).

Table 4.1 Hazardous-materials truck types.

Silhouette	**Description**
Low-pressure flammable liquid tank truck 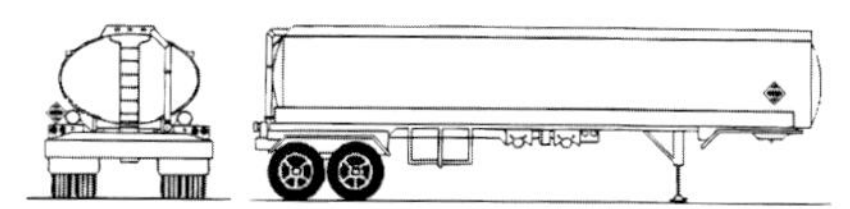	• Typically carries gasoline (UN/NA 1203), diesel fuel (fuel oil), liquid fuel products, alcohol, and almost any other kind of flammable or combustible liquids. May sometimes carry nonflammable liquids (e.g., milk or molasses). May contain mild corrosives, but not strong corrosives. Cannot contain pressurized gases. • Oval in cross-section, with blunt ends. • Newer tanks are aluminum; older ones can be steel. Tank is divided into two to five compartments (usually three to four); in some cases, different products may be in different compartments (in most states, mixed loads are not permitted). Typical maximum capacity: 34 000 l. Pressure can't exceed 20.7 kPa.
Low-pressure chemical tank truck 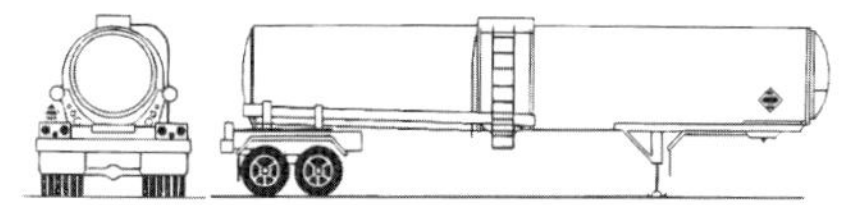	• Typically carries flammable or combustible liquids, acids, caustics, poisonous liquids. • Maximum capacity is typically up to 22 700 l. Pressure can be up to 276 kPa. Can be insulated or uninsulated: ○ Uninsulated tanks are typically circular in cross-section. Typically, there are reinforcing rings around the tank. Tanks are aluminum or steel. ○ Insulated tanks generally carry products that need to be kept either heated or cooled, or products that need to be heated to be off-loaded. They are characteristically horseshoe-shaped when viewed from behind. They are comprised of an outer jacket, generally aluminum or steel, and an inner tank that may be lined (e.g., with fiberglass).
Corrosive liquid tank truck 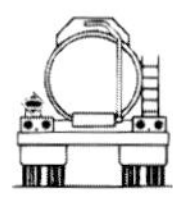 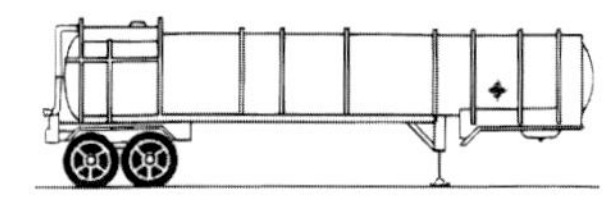	• Typically carries strong corrosives, such as sulfuric or nitric acid. Typically carries acids, also may carry bases. Sometimes may carry flammable liquids (e.g., grain alcohol), poison liquids, or oxidizing liquids. Cannot carry pressurized gases. • Circular in cross-section, with up to 10 reinforcing rings around the tank. May be very long. Often there is black, tar-like, corrosion-protective coating around the manhole. • Carries a single tank, generally with a single compartment, usually of steel and lined, with capacity up to 26 500 l. Tank pressures between 241 and 345 kPa.
High-pressure tank truck 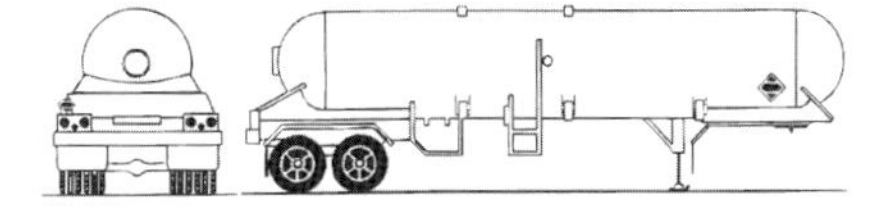	• Typically carries gases liquefied by pressure, such as anhydrous ammonia, LPG, propane, butane. • Circular in cross-section, with blunt ends: tank looks like a bullet. Surface is smooth; typically painted white or silver to reduce heating by sunlight. • Tank can carry up to 43 500 l; tank pressure is generally above 689 kPa. • Shorter "bobtail" versions can carry up to 13 250 l. • High BLEVE[a] potential.

(*Continued*)

Table 4.1 (*Continued*)

Silhouette	Description
Cryogenic liquid tank truck  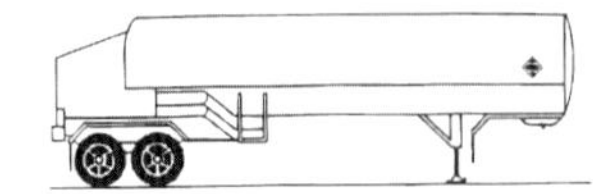	• Typically carries gases liquefied by refrigeration, such as liquid oxygen, nitrogen, argon, carbon dioxide, and hydrogen. Product likely to be corrosive or flammable gas, or poisonous or oxidizing liquid. Temperature of product −100 °C or below. • Outer shell surrounds insulated inner tank, with vacuum space between. Large compartment mounted at rear of tank. Capacity of inner tank up to 26 500 l. • When sun heats tank and raises internal pressure, vapor may discharge from relief valves. Internal pressure up to 172 kPa. • Very high BLEVE[a] potential.
Dry bulk cargo tank truck 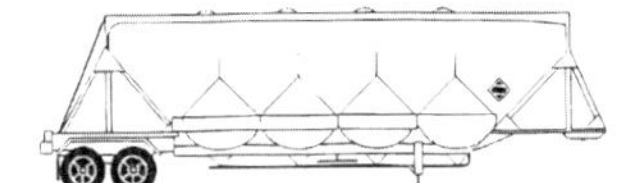	• Typically carries dry bulk cargo such as calcium carbide, oxidizers, corrosive solids, cement, plastic pellets, or fertilizers. • Shape can vary but always includes bottom hoppers.
Tube trailer (compressed gas trailer) 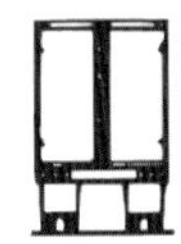 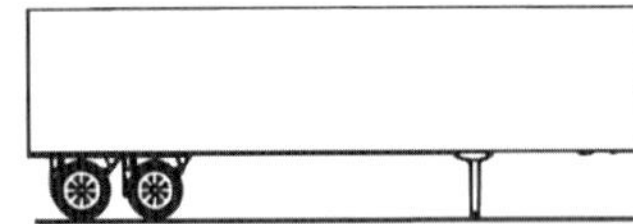	• Typically carries pressurized gases such as air, helium, and oxygen, in pressurized tubes. • Pressure may be up to 3400 kPa.
Box or van trailer (mixed cargo) 	• Typically contains mixed cargo, which may be packed in bags, boxes, drums, tanks, cylinders, or other containers. • The presence of several placard holders on the truck indicates it may commonly carry hazardous materials.

a) BLEVE = Boiling Liquid Expanding Vapor Explosion, which can occur when the material in a pressurized tank exceeds its boiling point and the tank is ruptured.

Source: NOAA Office of Response and Restoration for the U.S. Coast Guard, Chemical Response Tool *Field guide to containers & vehicles*. http://chemresponsetool.noaa.gov/Smart_Front_End.htm#Containers_guide/Truck.htm (accessed 18 February 2011).

4.3
Security-Sensitive Materials

While there may be many interrelated elements that a terrorist individual or organization may consider when selecting the weapon, the attack mode, and the target, the availability and specific nature of the materials to be used are certainly key factors. Initial efforts to address hazmat transportation security following the attacks of September 11, 2001 resulted in a broad application of rules to most carriers of hazardous materials, irrespective of their inherent hazards. In the current environment, a more varied approach has been adopted, where progressively increasing requirements or procedures are applied to materials deemed to pose the greatest hazard or risk. In the United States, for example, the Department of Homeland Security has developed different guidelines and recommendations for two tiers of highway security-sensitive materials (HSSM) (Transportation Security Administration, 2009a, 2009b).

Generally, the highest tier of highway security-sensitive hazardous materials includes Division 1.1–1.3 explosives; Division 2.3 toxic gases; certain Class 7 radioactive materials; and specified quantities of Division 2.2 nonflammable gases, Class 3 flammable liquids, and Division 6.1 poisonous materials that are also toxic inhalation hazards (TIH). The European Union does not distinguish between different tiers of security-sensitive materials and adds the following to their list of high-consequence dangerous goods: desensitized explosives, Division 4.1 flammable solids, Division 6.2 infectious substances, and specified quantities of Division 4.2 spontaneously combustible materials in Packing Group I, Division 5.1 oxidizing liquids of Packing Group I, and Class 8 corrosive substances in Packing Group I (European Commission, 2005).

Hazardous-waste shipments should also be considered as another category of material. In the United States, these shipments are regulated by the Environmental Protection Agency (requiring hazardous waste manifests) as well as transportation safety and security regulators. As hazardous waste has no economic value, there may be less concern placed on security and other scrutiny, such as timely reporting of late deliveries.

4.4
Carrier Responsibility

When considering road transport of hazardous materials, the primary focus is on the motor carrier or trucking company that will be moving the materials between the shipper and the consignee. This section focuses on carrier-focused security issues and Section 4.5 addresses shipper and consignee responsibility.

Motor carriers operate in an environment where there are strong financial incentives to minimize costs and maintain a healthy financial position. With the wide range in company size and resources, it is no surprise that there is a similar diversity in the focus, attention, and resources that different hazmat transportation companies direct toward improving their security posture. For some companies, carrying hazmat is the core of their business using specialized equipment and procedures and, for others, hazmat is a very small component of their operations.

In addition to internal company incentives to address security threats for economic, continuity, or even altruistic reasons, there are also federal and state regulations that are designed to ensure a consistent standard of safety and security across all carriers.

When addressing general hazardous-materials transportation security issues, the following broad categories are generally applied:

- General security issues, which includes an element of risk assessment and some consideration for varying the security posture based on evolving threat information
- Personnel security, which focuses on vetting personnel involved with hazardous-materials shipments in any way as well as protecting the safety of employees and contractors.
- Unauthorized access, which considers appropriate means of controlling access to facilities, systems, vehicles, and materials.
- *En route* security, which addresses the protection of shipments in transit.

Most countries require carriers of certain hazardous materials (such as high-consequence dangerous goods) to develop and maintain security plans that cover these broad categories. The following sections provide perspectives on how these issues are addressed in motor-carrier transportation of hazardous materials.

4.4.1
General Security Issues

Governmental agencies conduct transport risk assessments at national and regional levels to determine how to improve overall security, identifying appropriate approaches to address issues or deficiencies they uncover. At their disposal are a number of effective tools to achieve the security goals they adopt. These include guidance, training, enforcement, and regulation. While safety regulations have been established and evolved over a period of several decades, many security regulations are just now being considered and developed. A lot of the work to date in this area, particularly in the United States and Europe, has focused on providing guidance to industry on best practices to consider adopting. In some cases, regulations have been adopted to require security planning, highway route restrictions, and security training (European Commission, 2005; Federal Motor Carrier Safety Administration, 2003; Code of Federal Regulations, 2010).

Security Action Items (SAIs) are an evolving set of guidance provided by the United States Department of Homeland Security's Transportation Security

Administration. Carriers are encouraged to adopt as many of the SAIs as they can to improve their overall security posture. The SAIs are not mandatory and it is not expected that all carriers could or should adopt all of them. Contrast this with the situation with others such as in Singapore where the smaller scale supports centralized control and mandatory implementation of such security countermeasures as installing truck tracking and disabling technology.

Industry stakeholders in the hazmat trucking supply chain conduct risk assessments to support business decisions on the best alternatives to transport commercial products from their point of manufacture to their ultimate endpoint.

4.4.1.1 **Risk Assessment**

The risk-assessment process is designed to help prioritize and select security countermeasures that correspond to the threats, vulnerabilities, and consequences that may arise from a terrorist attack employing hazardous materials. One wishing to consider threat in an assessment without direct knowledge of terrorist intentions can consider the tiered approach used to designate varying levels of highway security-sensitive materials. One could also consider the frequency of shipments of varying types of materials to determine relative accessibility to terrorists. Vulnerability measures how well a given operation is susceptible to specific types of attacks. The consequences, of course, are determined by the impacts that may result from each type of attack. Consequences can include population, environmental damage, or economic effects. A proper assessment will consider these three elements together to determine the best approach to reducing risk. A certain type of attack may be very easy for a terrorist to carry out; however, low-potential consequences would offset this high vulnerability and make it much more likely that the adversary would choose another type of attack to pursue.

4.4.1.2 **Attack Profiles**

Terrorists generally are looking to maximize human health, economic, or emotional damage. Targets particularly attractive for road transports include large gatherings of people, iconic targets, and critical infrastructure. To a lesser extent, causing environmental damage may be an additional motive. Based on research by the US Department of Transportation, the most likely attack profiles against hazmat road transports are theft, interception and diversion, and legal exploitation (Federal Motor Carrier Safety Administration, 2003).

- Theft – taking the hazmat anywhere along the supply chain, from the shipper's location, from the motor vehicle at any point while *en route*, or during delivery at the consignee. Theft can occur quietly by insiders, or while materials are inappropriately left unattended, or forcibly.
- Interception – using the hazmat in an attack when the vehicle is at the target location during the normal course of transportation.
- Diversion – this is related to interception in that the hazmat shipment is diverted from its normal route so that it passes near to the intended target and

then used for an attack in that location. This can occur through the use of false detours, erroneous dispatch orders, or other means.

- Legal exploitation–simply purchasing the desired hazmat through normal commercial means and stockpiling sufficient quantities for an attack or intercepting shipments *en route* to the consignee location.

Theft and diversion are unique to road transport as described here, due to the vast extent of the roadway network and the opportunity to redirect materials from their normal shipping lanes to virtually any location.

Given the nature of many hazardous materials, particularly those deemed highway security-sensitive, they can be weaponized by a terrorist very quickly. There is little need to convert these materials into other forms or to repackage them in different ways to obtain an effective device. Flammable materials, for example, simply need a trigger to release the contents from the packaging (such as a cargo tank) and ignite it. This can be accomplished by improvised explosive devices (IEDs), vehicle-borne improvised explosive devices (VBIEDs), rocket-propelled grenades, and other means. TIH materials may be released by breaching the packaging or by opening or damaging valves or piping.

As mentioned earlier in this section, terrorists will try to align the attack mode, the material, and the target to best achieve their objectives. Some materials are more suited to different targets than others. For example, an attack using flammable liquids or explosives would be more effective on infrastructure than TIH materials. A recent study in the United States concluded that the most likely terrorist attack using hazardous materials would involve flammable liquids, such as gasoline, or secondarily, flammable gases, such as propane (Jenkins and Butterworth, 2010).

4.4.1.3 Planning an Attack

Terrorists have to successfully complete quite a few operational steps in advance of a planned attack (Federal Motor Carrier Safety Administration, 2003). Each of these provides an opportunity to interrupt the planning process and effectively prevent the attack. For road transport of hazmat, some of these operational steps may be more easily carried out.

As one example, casing and preattack surveillance of *en route* attacks on a hazmat vehicle may be conducted without raising suspicions as there is no protected area surrounding the road network. Cars on the road and parked alongside it, pedestrians, etc., can effectively observe passing vehicles with little concern, provided they are not adjacent to critical infrastructure that may employ security measures to detect or prevent such surveillance.

4.4.1.4 Training

One of the most important elements of security is training. Ensuring that all employees have security-awareness training and know what to do if something occurs is critical to maintaining a solid security posture. Furthermore, adequate training may even allow employees to prevent or deter a terrorist attack, due to increased awareness and heightened alert. Periodic testing of employees can be an effective tool to reinforce the training and refine procedures over time.

4.4.2
Personnel Security

The foundational element of personnel security is proper and thorough vetting of all personnel that come into contact with or can influence a shipment of hazardous materials. Personnel security begins with the job application and hiring process and extends through to employee termination. In motor carrier transportation, the driver is often the focus of attention, but dispatchers, facility and dock workers, clerks, and security personnel are all key positions that require security considerations.

Drivers have the greatest opportunity to subvert security measures as they are alone with the hazardous materials throughout the entire shipment. Drivers in the United States are required to obtain a hazardous-materials endorsement to their commercial driver's licenses, which includes a Security Threat Assessment conducted by the Transportation Security Administration to specifically address this vulnerability. Depending on the nature of their operation, drivers may also need a number of other credentials (see Table 4.2), all of which include some degree of vetting for security purposes.

Internationally, similar situations apply. For example, the European Agreement concerning the International Carriage of Dangerous Goods by Road (ADR) specifies that drivers of vehicles carrying dangerous goods must hold a certificate issued by the competent authority or by an organization recognized by that authority stating that they have participated in a training course and passed an examination on the particular requirements that have to be met during carriage of dangerous goods. In countries like Singapore, all drivers must have a Hazmat Transport Driver Permit (HTDP), which includes driver security screening and a requirement to pass an exam (following a one-day class) on hazmat transportation safety.

Of course, drivers are also a key component of the security of the shipment and, when properly trained, can adjust to changing conditions and apparent threats, reporting them to the proper authorities if something seems amiss. Driver turnover can make it more difficult to maintain thorough vetting of potential new hires and most carriers prohibit less experienced drivers from transporting hazardous materials.

Dispatchers may also pose a risk of insider collusion. They have the ability to redirect *en route* shipments to other locations, whether to a target or on a route that will pass close to a target.

All individuals involved in the supply chain can be a source of information for a terrorist planning an operation. Inadvertent disclosure of sensitive information useful to a terrorist can be easy to achieve.

Table 4.2 US driver credentials.

Credential	Description
Hazardous Materials Endorsement (HME) to a Commercial Driver's License (CDL)	CDLs issued by state licensing authority; HME vetted by the Transportation Security Administration. Required to operate a commercial motor vehicle carrying hazardous materials.
Transport Worker Identification Credential (TWIC)	Issued by the Transportation Security Administration for everyone requiring unescorted access to secure areas of certain facilities and vessels.
Free and Secure Trade Card (FAST), NEXUS, and Secure Electronic Network for Travelers Rapid Inspection (SENTRI)	Issued by Customs and Border Protection to facilitate crossborder travel between the US and Canada and Mexico.
US Passport	Issued by the State Department and required for re-entry into the US
Common Access Card	Issued by the Department of Defense for access to military facilities.
Port, airport, and other facility access cards	Issued by the respective entities for access to security areas.

Personnel security can also refer to protection of employees. This aspect includes some elements of unauthorized access, discussed in the next subsection, as well as policies and procedures to report suspicious or potentially dangerous behavior. Of course, plans must be in place for quick response to reports of suspicious behavior.

4.4.3
Unauthorized Access

There are numerous opportunities for unauthorized persons to gain access to hazardous materials shipped by road. It is important to maintain a secure chain of custody for each shipment. Some of the key opportunities are listed here:

- Carrier facilities–while most of the personnel at a carrier facility will be company employees, there should be procedures in place to properly badge, escort, or announce visitors, contractors, and vendors. At larger facilities, proper badging of all employees should include photo identification and procedures for challenging and addressing persons without proper badges. Many facilities are fenced with control access, but plenty of smaller operations allow easy access inside throughout the day with the assumption that onsite staff will observe and question any unexpected visitors.

- Loading docks – it is important to ensure that each shipment is loaded on the proper vehicle and that all materials awaiting loading are adequately secured and monitored.
- Storage facilities – particularly important for intermediate storage or warehousing.
- Consignee/unloading facilities – staff here should be vetted as well to ensure that arriving materials are not diverted to other vehicles or locations.
- Dispatch systems – preventing inappropriate shipments or diversions of legitimate shipments is the focus of access security of these systems.
- Inventory systems – controlling the quantities of materials at all points throughout the supply chain can have a great impact on determining when theft has occurred. Access into an inventory system (manual or electronic) can give a terrorist an opportunity to hide their tracks.
- Other technology systems – these can range from vehicle-routing and cargo-tracking tools to employee and approved contractor records.
- Vehicles – preventing unauthorized access to vehicles is a critical element. Simple procedures from always locking vehicles when unattended can go a long way. Some firms are using biometric identification to uniquely recognize the assigned driver before allowing the vehicle ignition to function.
- Cargo – a lot of technology has been applied to protecting the cargo from unauthorized access. Mechanical, label, and electronic cargo seals are becoming more widespread, but they cannot be effectively applied to all types of packagings. Electronic seals tied in to communication systems (such as those often installed in-cab for driver–dispatch communications) can automatically report unauthorized access to the cargo, such as when the vehicle is outside the origin or destination location. Seal-custody programs are critical to the successful use of cargo seals (Cook and Farrar, 2007). There are also issues regarding enforcement actions and securely reapplying cargo seals when they are broken during an official inspection.

For the electronic systems, it is important to ensure that appropriate access is maintained for direct connections, such as from a corporate computer onsite, and for remote connections, such as from a driver's mobile device or by using a company official's login credentials. Understanding the appropriate security classification and allowed users for certain information is critical to maintaining effective security.

One good example is dispatch authority. In some companies, only dispatchers assigned to a particular facility are allowed by the system to make adjustments to manifests of shipments that originate from that facility. This effectively precludes a coopted dispatcher at another facility from affecting shipments with which they should not be interacting.

4.4.4 En Route Security

By far the greatest vulnerability to road transport of hazardous materials is when the shipment is actually on the road. Unlike at a facility, there is no intermediate protective zone around the vehicle and its cargo other than the packaging itself. Many shipments of security-sensitive materials are either not allowed to stop *en route* or have regulatory or policy constraints on their on-road behavior. The section below offers a discussion on many of the security issues a motor carrier faces while *en route*.

- Packaging selection – most often it is the shippers that determine the packaging used for a given shipment and security considerations may apply. More robust packaging may be less vulnerable to sabotage or some standoff weapons (e.g., high-caliber rifle), but in general, security concerns do not seem to play a significant role in packaging selection between various alternatives.
- Known shippers and attended loading – when dealing with sensitive materials, a carrier may consider it a security risk to transport materials from shippers that it has not thoroughly vetted. These programs are often referred to as "known shipper" programs. In some cases, shippers may offer a carrier a previously loaded container or trailer. It is prudent to have carrier personnel attend the loading process, even if performed by the shipper's employees. This will help ensure the proper materials are transported and that they are packaged safely and securely.
- Route selection – keeping security-sensitive hazmat shipments away from potential targets can be an effective countermeasure. In some jurisdictions, local and state authorities have the power to implement route restrictions on certain hazardous-materials shipments. They can consider both safety and security issues in determining whether to preclude some or all hazmat shipments from taking certain roadways or requiring that they take certain roadways if traveling along a prescribed route (such as a beltway if traveling through an urban area). Shippers and carriers may explore route alternatives for the same reasons – to reduce their overall risk – and security considerations can be important. Some shippers and carriers have expressed a willingness to take less economical routes in exchange for reducing the safety and security risks. Specific avoidance of high population concentrations (even intermittent ones, such as sporting events) and critical infrastructure are the two most critical considerations. Bridges and tunnels are particularly important; especially if there are limited alternative routes should they be rendered unusable after an attack. Also note that some materials require written route plans; these include Highway Route Controlled Quantities (HRCQ) of radioactive materials and

Division 1.1–1.3 explosives in the United States (Code of Federal Regulations, 2010). One issue that can be overlooked when considering alternate routing to avoid high-population areas is the level and response time of qualified hazardous materials emergency response teams. These teams are more often concentrated in the high-population areas and response times to more remote areas – assuming they contain other attributes that are attractive to terrorists – can be greater than desired.

- Unattended shipments – some shipments are prohibited from being left unattended, but for those that can be left – even briefly – there is a considerable security risk. Drivers returning to vehicles left unattended should thoroughly inspect their vehicles for tampering and affixed explosive or unfamiliar or foreign tracking devices.
- Unattended loading – In some cases, notably fuel transportation, drivers may make unattended pickups where they have access cards and codes that allow entry into a tank farm or other loading area where no shipper personnel are present.
- Truck stops – the presence of large numbers of people familiar with road-transport operations offers some security through the increased awareness of those around any commercial motor vehicle. Some truck stops offer increased security as a marketing feature to attract business. However, drivers should be aware of leaving their vehicles unattended even in these locations. In addition to technology solutions discussed later in this section, drivers should ensure that their vehicle, trailer, and cargo can be securely locked (see Table 4.3). Safe havens are a concept where locations with adequate security are designated as acceptable *en route* locations in the event of circumstances that prevent a shipment from continuing on to its destination. Cargo is generally allowed to be left unattended if parked in a safe haven.

Table 4.3 Truck and trailer locking devices.

Device	Description
Trailer Locks	Standard locks that secure the cargo area of the trailer
Tamper Evident Seals	May not prevent theft, but provide obvious indication of tampering.
Trailer King Pin Lock	Fits over the king pin and prevents the trailer from being connected to a tractor.
Air Brake Glad Hand Lock	Prevents moving the trailer by blocking the air intake valve of the emergency brake system.
Parking Brake Locks	Similar to glad hand locks, they fit over the air intake valves of the parking-brake system.

- Rest/parking areas – much less secure than a truck stop, rest and parking areas offer little protection, little opportunity for others to watch out for suspicious behavior, and often a fairly isolated location far from potential response to any incident that might occur.
- Inspection and weigh stations – these may provide a level of protection when law-enforcement officers are present if the driver needs to take a required rest break.
- Tandem drivers – where longer distances do not allow a single driver to complete the route without a rest period, tandem drivers are often used to expeditiously deliver hazardous materials and ensure continuous monitoring of the vehicle and its surroundings.
- Escort vehicles can be effective countermeasures and may be required for certain materials in certain locations. Escorts could be armed or unarmed and potentially trained in emergency-response procedures; they would be aware of the materials they were escorting. Escorts could include law-enforcement officers as well as company and contract employees.

4.4.4.1 Driver Responsibilities

Driver pretrip and *en route* walk-around inspections are an important element in hazmat road transport. These inspections should be conducted any time the driver is away from the vehicle. Drivers should be looking for evidence of tampering with valves, seals, locks, pipes, and other safety and security equipment. They should also look for the introduction of foreign objects that may present a hazard – including explosive or tracking devices. Some carriers do not permit drivers to climb aboard cargo tanks to visually inspect the tops of the tanks for personal safety reasons, but that may provide an opportunity for a terrorist to plant a device on the vehicle.

Drivers must be entrusted and encouraged to report suspicious behavior and conditions while *en route*. Detailed policies and procedures need to support such reporting. Drivers, certainly those transporting security-sensitive hazardous materials, should be extra vigilant about seemingly innocuous but potentially dangerous situations, such as solicitations for aid from other motorists.

4.4.4.2 Technology Solutions

Communication equipment is very common today on most vehicles carrying hazardous materials. Most drivers carry at least a mobile phone; whereas, many companies use some combination of cellular and satellite system to ensure greater coverage. Communication policies and procedures allow the dispatcher to maintain some level of awareness as to the current status and location of the driver.

Tracking technology is also fairly widely deployed across some segments of the hazmat commercial motor carrier industry. Some short-haul operations, such as gasoline distribution, tend to not deploy tracking technology as they are regularly in contact with their drivers as they make their frequent deliveries throughout the day. While most tracking is limited to the truck or tractor (for combination vehicles), there are trailer tracking systems as well. These are often used in addition to truck-tracking solutions and can provide position information for untethered trailers because of self-contained battery or power supplies.

Attendant with some communication or tracking systems is driver alert or "panic button" functionality. This can be connected to an in-cab communication device or retained by the driver on a key fob. In either case, the driver can send a message immediately to their dispatch operations to notify them of an emergency situation. This can be a security, safety, or medical issue, but until the dispatcher can communicate with the driver to determine the exact nature of the emergency, they must assume it is a security incident.

Other technology employed includes:

- Biometric identification and authentication to ensure that the driver is authorized to operate the specific vehicle at that time. Fingerprint readers are the most commonly used technology but its use is not widespread.
- Electronic cargo seals present another opportunity for technology to impact security.
- Vehicle immobilization is beginning to get greater consideration for adoption in many parts of the world. Serious cargo-theft problems in some areas have helped advance the technology for and adoption of remote vehicle immobilization (Palmer, 2010). In fact, vehicle immobilization refers to two different types of technologies. Vehicle disabling generally refers to preventing a vehicle from starting. Vehicle shutdown is used to stop a moving vehicle and is usually implemented in a manner that allows the vehicle to limp to the shoulder of the roadway safely by limiting acceleration or reducing the throttle or engine power. Vehicle immobilization can be activated by the driver in cab by entering an invalid or duress code into an onboard computer or by pressing a remote panic button. Vehicle immobilization can also be initiated by the dispatcher when he or she is concerned about a potential or confirmed security incident. Some disabling technology can be triggered when tracking or other systems are themselves disabled. (Franzese *et al.*, 2008)
- Geofencing is a tool where the position information obtained from vehicle- or cargo-tracking equipment is compared to predefined spatial areas and exceptions are flagged for warning or action. Geofences can be structured as inclusions or exclusions. For inclusions, the allowed areas for the vehicle are specified, such as along a predefined route path. Allowances are made for regular excursions for fuel or other planned stops. For exclusions, areas where the vehicle is not allowed to go are defined. These can be protected areas around critical infrastructure, high population areas, iconic targets, etc.
- Remote cargo-door locks require the dispatcher to use a communication system to lock or unlock the trailer door, typically upon driver notification. Location information can be coupled with this technology to enforce security

protocols such as only allowing the doors to be unlocked at an origin or destination facility. Special cases would be need additional authorization, such as when a roadside inspector requests to view the cargo to check for regulatory compliance with packaging and shipping requirements.

- Cargo monitoring refers to installing sensors to detect unintentional releases of hazardous materials from their packaging, for example, from the valves on a cargo tank. Such sensors are usually specific to the material being transported and are often installed for safety reasons, but the security benefits are readily apparent. Sensors can report information to the driver via direct or wireless communications and may also relay information to a corporate dispatch center. Communication with a dispatch center can occur via cellular or satellite directly using equipment integrated with the sensors or by relaying the information through the in-cab communications systems.

Electronic shipping papers are used in other modes such as rail and are being considered for highway transportation as well. They would eliminate problems that may arise when physical shipping papers are unavailable after an incident and to support both motor-carrier enforcement and emergency-response operations. The security implication for this technology relates to keeping the supply chain secure and allowing access to the information when it is needed by authorized personnel. Added benefits include reduced costs to those that are already using electronic communications in their supply chain. Implementing electronic shipping papers could involve two different approaches, either separately or in combination: (i) Internet-connected communication devices accessing databases containing the manifest information or (ii) radio frequency identification (RFID) tags that contain the information. In the Internet-based approach, some identifier such as the license plate, tractor number, or trailer number would be needed to uniquely look up the shipment. RFID data would be located with the vehicle and would require a secure reader to allow access to response personnel and not unauthorized individuals. Some of the same incidents that would render printed shipping papers unavailable might also damage the RFID tags.

The HazMat Transport Vehicle Tracking System (HTVTS) introduced in Singapore by their Civil Defense Force (SCDF) in 2005 is a good example of a national system applied to hazmat road-transport security, albeit in a unique environment. Singapore has only one level of government and a limited number of hazardous-materials shippers and carriers. This system can accept data from varying technologies including global positioning systems (GPS), short message systems (SMS), general packet radio services (GPRS), and inertial navigation systems (INS) to track hazardous-material vehicle locations. The system also allows monitoring of driver identity and hazmat inventory.

The Singapore HTVTS is setup to alert SCDF staff when the following conditions occur:

- the tracking device is tampered with;
- a vehicle diverts from an approved route (implemented via a geofence);

- a vehicle moves outside approved timeframes;
- a vehicle enters a exclusionary geofence;
- a trailer is untethered without authorization.

In Singapore's particularly small, high-population urban environment, the SCDF determined that there was insufficient time to react to a route deviation with response personnel, so they developed a vehicle immobilization system in 2008 (Seng, 2008).

4.4.4.3 Motor-Carrier Enforcement

Motor-carrier enforcement can also be a motivator for companies to implement and maintain effective security programs. Hazmat carriers are subject to additional inspections at the roadside to ensure that the materials being transported are consistent with safety and security regulations. Accidents involving hazmat carriers may affect their ability to continue to carry certain materials.[4)]

4.5 Shipper and Consignee Responsibility

When discussing road transport of hazardous materials, the primary focus is on the motor carrier. The entire supply chain needs to be included, however, when considering security issues. As is the case with other transport modes, many manufacturers and other shippers select their preferred carriers based, in part, on their security programs. Shippers should implement systems to ensure that the correct driver picks up the correct cargo.

Consignees play an important role in supply-chain security as well in that they are often involved with the unloading process and, in addition to personnel security issues, they must maintain a rigorous inventory system to ensure that the amount received matches the amount shipped and that discrepancies are immediately reported and researched. They must also implement systems to ensure that the right products are being delivered to their locations and by trusted drivers.

4.6 Motor-Carrier Enforcement

As discussed in Section 4.4.4, motor-carrier enforcement can play a big role in improving security as well as safety. First, roadside inspections may provide enforcement officers the opportunity to identify potentially unauthorized drivers (through their interactions and observation of suspicious behavior), to ensure that the material is packaged in a safe and secure manner, and that there are no suspicious devices placed in or on the vehicle. In addition, compliance reviews or

4) FMCSA Safety Permit Program www.fmcsa.dot.gov (accessed 28 February 2011).

security visits to corporate headquarters or loading or unloading facilities may uncover other safety concerns that warrant resolution.

4.7
Law Enforcement and Emergency Response

Proximity to law-enforcement and emergency-response personnel is an important element in road transportation security, particularly where hazardous materials are involved. If a driver, dispatcher, automated system, or concerned observer reports suspicious behavior involving a truck carrying hazardous materials, being able to notify the appropriate law-enforcement authorities to initiate a timely response is very important. Many dispatchers maintain lists of emergency contacts throughout the routes their drivers take. Universal emergency numbers (such as "911" in the United States and Canada) can assist where available. If the incident, such as an attempted theft of the vehicle, occurs far from populated areas where the law-enforcement personnel are concentrated, it may be easier for a terrorist to obtain the materials and move them to their target or transload them into another vehicle.

If a terrorist is successful in completing an attack involving a truck carrying hazardous materials, timely emergency response can be very effective in limiting the consequences of some types of attacks. If radiological materials or biological agents are involved, it is even more critical that qualified emergency responders arrive on the scene to identify the nature of the incident and take the appropriate precautions to limit the extent of the consequences. As a truck-involved hazardous-materials security incident may have occurred on a major roadway, evacuation and response routes may be affected, further exacerbating the situation. Ideally, local emergency-response planners have considered the likelihood of a terrorist attack in their community and have prepared appropriate contingency plans.

4.8
Community Vigilance

Unlike other modes of transport, road transport of hazmat is conducted in close proximity to the general public and other transportation-sector personnel, providing them an opportunity to observe and notice suspicious behavior or other security-related concerns. There have been a number of efforts in the United States to educate others on specific highway- and hazmat-security awareness. Currently, the First Observer program, operated under a cooperative agreement with the Department of Homeland Security's Trucking Security Program, provides training in the trucking, school bus, and motor coach carrier sectors on antiterrorism and security awareness[5]. Past efforts have provided training to toll booth operators, roadway maintenance personnel, and others.

5) www.firstobserver.com (accessed 27 February 2011).

4.9 Security-Related Events

History provides some concrete examples of terrorist attacks on commercially transported hazardous materials by road. An April 2002 attack in Tunisia involved a liquefied natural gas (LNG) shipment that was ignited near a synagogue, killing 21 people and injuring 30. Also in 2002, unsuccessful attacks involving placing remotely controlled IEDs on fuel tankers were attempted in Israel.

In December 2003, the Department of Homeland Security notified hazardous materials trucking industry officials of "possible indicators of terrorist attack planning." The notice reported five incidents in three states in which fuel truck drivers reported seeing "Middle Eastern men" conducting "surveillance."

Beginning in 2007, Iraqi insurgents began to attack TIH-carrying trucks in high-population areas, releasing such gases as chlorine. Two particular attacks killed 5 people and injured more than 200 (Seng, 2008).

Insider threats are real in the road transport of hazardous materials. An al Qaeda-connected plot intended to destroy the Brooklyn Bridge and cause train derailments near Washington, D.C. involved a truck driver authorized to transport hazardous materials named Iyman Feris (United States Department of Justice, 2003; Schumer, 2004). He was convicted in 2003.

4.10 Conclusions

Road transport of hazardous materials accounts for the largest share of shipments by far (at least in the United States and presumably most of Europe and a majority the world) (Bureau of Transportation Statistics, 2010; European Union, Eurostat. http://epp.eurostat.ec.europa.eu/portal/page/portal/transport/data/database, Modal Split of Freight Transport table (accessed 18 April 2011)). Increasing terrorist use worldwide of commercial hazardous-materials shipments as weapons in attacks will continue to influence the already heightened awareness of the risks to road transport of these materials.

As resources are limited, a greater emphasis will be placed on applying appropriate technologies to address the various vulnerabilities in the supply chain. Advanced monitoring techniques and centralized truck tracking will support large-scale situational awareness concerning these shipments. Packaging designs, such as for cargo tanks, are also continually evolving and will become more and more robust for safety reasons, but will be adding additional security protection as well. Routing to reduce security risk is also likely to be adopted more often in the future. Considering the proximity of hazardous-materials road transport to critical infrastructure and high population densities in routing decisions may be required by regulators.

Care should be taken to ensure that the smallest operators are brought along as advances in security techniques and understanding of the threat evolves – terrorists will seek out the weakest links in the targeted industry to improve the success of their attacks. The largest carriers are traditionally the ones that have sufficient resources to direct to security-related efforts and risk-mitigation strategies. Today, the trucking industry is a safe and secure one today, but vulnerabilities do exist and improvements can always be made.

Bibliography

Bureau of Transportation Statistics (2010) 2007 Commodity Flow Survey, Hazardous Materials Series, U.S. Census Bureau, Washington, DC.

Code of Federal Regulations (October 2010) Title 49, Part 397 Subparts C and D, United States Code, United States of America.

Code of Federal Regulations (October 2010) Title 49, Chapter III, Subchapter B, United States Government.

Cook, G.R. and Farrar, M. (2007) User's Guide on Security Seals for Domestic Cargo. United States Department of Homeland Security.

European Commission (2005) Industry Guidelines for the Security of the Transport of Dangerous Goods by Road, http://ec.europa.eu/transport/road_safety/topics/dangerous_goods/index_en.htm (accessed 18 April 2011).

Federal Motor Carrier Safety Administration (2003) *Guide to Developing an Effective Security Plan for the Highway Transportation of Hazardous Materials*. United States Department of Transportation.

Franzese, O., DeLorenzo, J., Knee, H.E., and Urbanki, T. (2008) *Vehicle Immobilization Technologies: Identification of Best Practices*. Presented at the Transportation Research Board 87th Annual Meeting.

Jenkins, B.M. and Butterworth, B.R. (2010) Potential Terrorist Uses of Highway-borne Hazardous Materials. MTI Report 09-03. Mineta Transportation Institute, San José, CA.

Palmer, J.S. (2010) *The Cargo* Theft Threat. InboundLogistics.com. http://www.inboundlogistics.com/articles/features/0110_feature10.shtml (accessed 27 February 2011).

Senator Schumer, C.E. (2004) New FBI Warning about Truck Bombs at High-Profile NYC Sites Shows Need for Comprehensive Anti-Terror Truck-Bomb Plan.

Seng, S.C. (2008) Safety and Security System for Hazmat Transport Vehicles in Singapore, Asian Disaster Reduction Conter March 2008.

Transportation Security Administration (2009a) *Security Action Items, Appendix A – Description of Voluntary Security Action Items for Tier 1 Highway Security-Sensitive Materials (Tier 1 HSSM) and Tier 2 Highway Security-Sensitive Materials (Tier 2 HSSM)*. United States Department of Homeland Security.

Transportation Security Administration (2009b) *Security Action Items, Appendix B – List of Tier 1 Highway Security-Sensitive Materials (Tier 1 HSSM) and Tier 2 Highway Security-Sensitive Materials (Tier 2 HSSM) with Corresponding Security Action Items*. United States Department of Homeland Security.

United States Department of Justice (2003) Iyman Faris Sentenced for Providing Material Support to al Qaeda, http://www.justice.gov/opa/pr/2003/October/03_crm_589.htm (accessed 18 April 2011).

USA Today (2003) Airport travelers urged to keep eyes open, http://www.usatoday.com/news/nation/2003-12-19-stay-alert_x.htm (accessed 18 April 2011).

5
Security Aspects of Hazmat Transport Using Railroad

Manish Verma and Vedat Verter

5.1
Introduction

> May 28th, 2010: At least 65 people were killed and 200 injured when a bomb apparently planted by Maoist rebels derailed an overnight passenger train in India, causing it to crash into an oncoming cargo train.

> March 29th, 2010: At least 38 people were killed and more than 60 injured in two suicide attacks on the Moscow Metro during the morning rush hour.

What are the similarities in the above events? *First*, both are forms of terrorist attacks targeting transportation system. *Secondly*, the underlying objective in both was to inflict maximum damage. Both these characteristics were also evident in the 2006 co-ordinated bombing of the Mumbai suburban railway, the 2005 serial suicide attacks on the London public-transport system, and the 2004 detonation of the Madrid commuter-train system. Although not all terrorist attacks would necessarily involve transportation systems, almost all are aimed at noncombatants with the objectives of instilling fear and causing catastrophic consequence.

The National Counterterrorism Center, founded in 2004 by the United States Government, is mandated with the task of collecting every recorded terrorism event across the world.[1] Over the twenty-year period for which data was analyzed, a total of 52 279 terrorist attacks took place claiming around 43 000 lives. It was also clear that transportation systems make attractive targets of intentional attacks since they accounted for 34% of the total attacks over the indicated period, and increased to 58% post 1998 (Leung, Lambert, and Mosenthal, 2004). On additional analysis, it was clear that vehicles (i.e. trucks, hazmat tankers, cars, etc.) were targeted the most, while air and marine modes were targeted the least (Figure 5.1). Although trains and buses account for 6% of the attacks, most of them involved the latter mode (Table 5.1). Fortunately, there have been far fewer episodes of terrorist attacks on trains/subways, but just like the two events outlined earlier – the

1) National Counterterrorism Center, www.nctc.gov (accessed 15 March 2011).

Security Aspects of Uni- and Multimodal Hazmat Transportation Systems, First Edition. Edited by Genserik L.L. Reniers, Luca Zamparini.

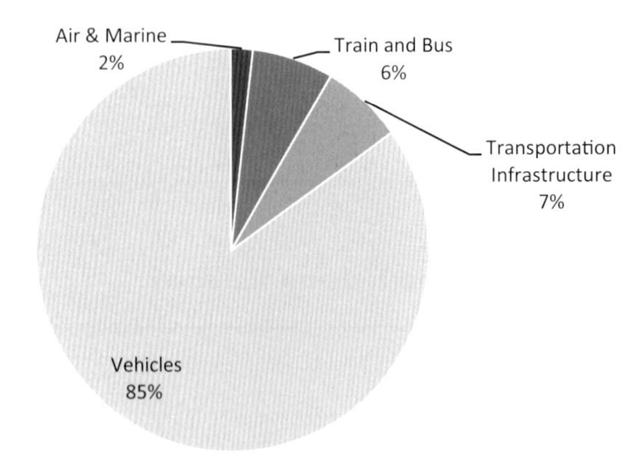

Figure 5.1 Proportion of attacks.

Table 5.1 Number of attacks and facility type.

Facility type	2001–05	2006	2007	2008	2009	2010
Aircraft & airport	66	20	37	47	43	25
Bus	279	213	160	132	104	72
Ship	23	7	12	20	17	4
Train/Subway	42	44	55	27	37	31
Transportation infrastructure	259	243	234	171	182	114
Vehicles	2448	2470	2670	2653	2810	2225

results have always been devastating. In addition, the use of transportation mode as both a target and a weapon, on 11th September 2001, has forced the government and railroad industry in North America to conduct appropriate vulnerability assessment, and then develop counterterrorist plans. This was important since the railroad industry not only operates a substantially more visible (and vulnerable) infrastructure that frequently moves through congested areas, but also because it transports a significant volume of hazardous materials.

Hazardous materials (hazmat) are harmful to humans and the environment because of their toxic ingredients, but their transportation is essential to sustain our industrial lifestyle. A significant majority of hazmat shipments are moved via the highway and railroad networks. In the United States, railroad carries approximately 1.8 million tons of hazmat annually, which translates into 5% of rail freight traffic.[2] On the other hand, in Canada, approximately 500 000 carloads of

2) Association of American Railroads *Current Rail Hazmat Conditions Called "Untenable"*, AAR New Press, http://www.aar.org/ViewContent.asp?Content_ID=3763 (accessed 15 December 2010).

hazmat – equivalent to 12% of total traffic – are shipped by railroad (Transportation Safety Board of Canada, 2004). The quantity of hazmat traffic on railroad networks is expected to increase significantly over the next decade, given the phenomenal growth of intermodal transportation and the growing use of rail–truck combination to move chemicals. Fortunately, a host of industry initiatives and the implementation of a comprehensive safety plan of the Federal Railroad Administration makes railroads one of the safest modes to transport hazmat (Federal Railroad Administration, 2008). In spite of the favorable safety statistic of railroads (Oggero *et al.*, 2006), the possibility of catastrophic events – however small – does exist. Although a number of efforts – over the past two decades – have been made to approach *catastrophe* associated with hazmat transportation from a "safety" perspective, the post-9/11 era also calls for investigating the "security" aspects of hazmat shipments (i.e. using them as weapons or targets).

This chapter is a first attempt to study the "security" aspects of railroad transportation of hazmat. To that end, we first describe freight railroad transportation in Section 5.2, where we also highlight the importance of "security". Section 5.3 contains a discussion on risk assessment from a security perspective, and outlines a methodology to estimate and/or determine different risk parameters. Section 5.4 outlines the different risk-mitigation techniques, including routing and interdiction strategies, to manage terrorist risk. Finally, Section 5.5 contains the conclusion and directions for future research.

5.2
Railroad Transportation System

The railroad industry – crucial to the United States and Canadian economy – has long considered itself to be the safest and the most secure mode of transportation, as well as the most efficient. As owners of rights-of-way, structures, and equipment, railroads have long invested in safety measures, including employing significant numbers of police officers and special agents, to guard their property (Plant, 2004). In the United States railroads have carried hazmat for at least the past century, and the current estimate of the number of carloads is 1.7–1.8 million.

To get a sense of the enormity of the problem, consider Figure 5.2 that depicts the freight network of a single railroad company, and is composed of rail yards and tracks. Some of the yards are fully equipped, that is, both classification and transfer operations are possible, while others can only perform block-swap functions (i.e. the hollow nodes). Any two nodes are connected by tracks, which are the service legs of a freight train traveling nonstop between them. A sequence of service legs and intermediate yards constitutes an itinerary available to a railcar for its journey. For major freight railroads, demand is expressed as a set of individual railcars that share a common origin and destination yard. To prevent railcars from being handled at every intermediate yard, railroads group several railcars together to form a *block*. A block is associated with an origin–destination pair that may or may not be the origin or destination of the railcars contained in the block.

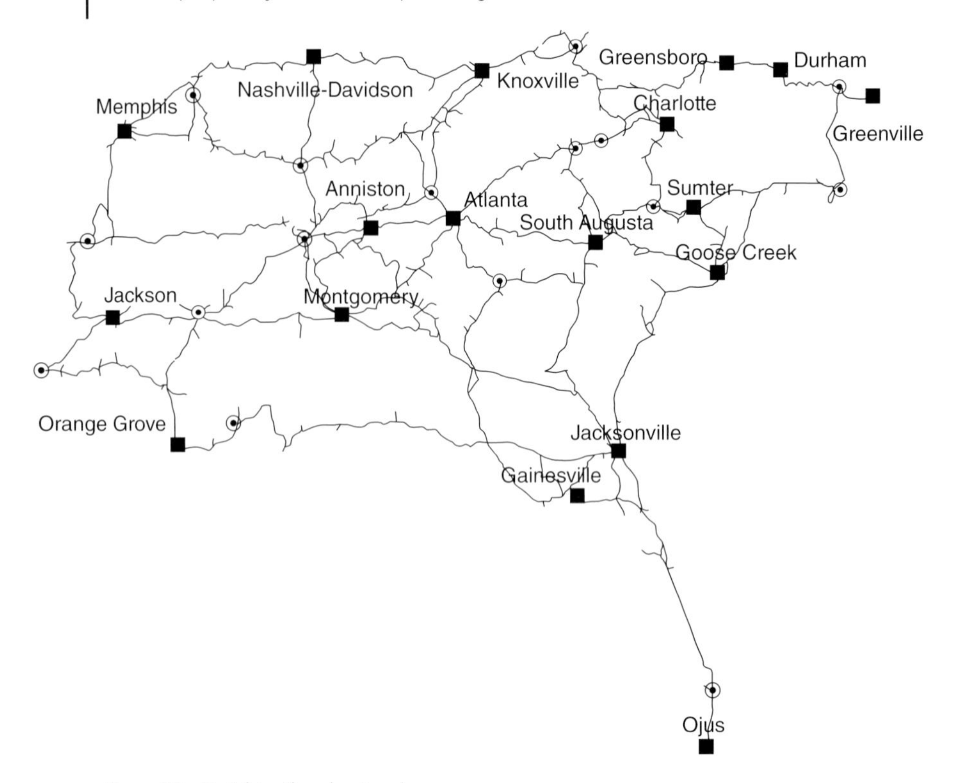

Figure 5.2 Freight railroad network.

The sequence of blocks to which a railcar is assigned on its journey from the origin to the destination yard is called a *blocking path* (Barnhart, Jin, and Vance, 2000).

Pre-9/11, the guiding policy of the federal program and the railroad industry was "safety" and not security, and hence the mission was to prevent fatalities, injuries, and property damage related to railroad operations and releases of hazmat from railcars in the respective freight networks (Citizen's for Rail Safety, 2007). But post-9/11, the two agencies have realized the security dimension of hazmat railcars–that is, a terrorist could either target them or use them as a potential weapon, and inflict spectacular damage on the population, environment and the infrastructure (Milazzo *et al.*, 2009). Conceivably, every yard and each track section could be a possible entry point for an individual with evil intentions, and hence protecting the railroad infrastructure would be daunting. This is not only because of the geographic dispersion of the railroad infrastructure, but also due to the nature of thinking required to assess the impact of newer aspects of rail operations (such as intermodalism, just-in-time deliveries), and the absence of enough empirical data to understand the mindset of terrorists. Given the spate of terrorist attacks on (passenger) trains overseas, it is reasonable to be concerned that the enormity of the physical and communications infrastructure could potentially be exploited by a determined adversary with potentially catastrophic consequences. The importance of appropriate planning is further underlined in the RAND Corp. database that depicts over 250 terrorist attacks against rail targets from 1995 to 2005 across the globe (Hartong, Goel, and Wijesekera, 2008).

5.3 Risk Assessment

Fortunately, terror events are extremely rare, but they are almost always aimed at noncombatants and used to instill fear in the targeted population. Although absence of enough empirical dataset does preclude experiential learning, one could conceivably work backwards from the end objective of any terrorist attack, namely catastrophic consequence. Translated to a threat involving railroad transportation of hazmat, this would seem to be equivalent to using one or more hazmat railcars as a weapon (or a target) with devastating results, near or close to a major urban center(s). For instance, every year over eight thousand tank cars of chlorine move by rail within two blocks of the US capital, and successful targeting of even one of these could potentially kill or injure hundreds of thousands. Similarly, an attacker who gains control of a hazmat tank car – in a rail-yard or in geographically dispersed and often lightly guarded industrial sidings – could use it as a weapon.

The vast railroad infrastructure and the existence of multiple railroad companies, each with its own network, calls for a cooperative development of a structured and systematic methodology to manage terrorism risk. Although it was reported that the railroad industry and the government agencies conducted a network-wide assessment to identify risk, and update emergency preparedness, neither the study nor the results are publicly available (Plant, 2004; Citizen's for Rail Safety, 2007). But in general, the first step in such a methodology should be to identify, quantify and measure risk.

The first step is to identify the sources of risk, that is, potential weak links/soft spots in the network. In general, potential vulnerabilities of railroad can be grouped under physical and communication infrastructure (Hartong, Goel, and Wijesekera, 2008), although it would also be important to have an idea (or intelligent estimates) about various (conditional) probabilities and consequences resulting from possible intrusion in the network. To that end, a risk filtering, ranking and management framework has been proposed to assess the threat of terrorist attacks against (highway) bridges (Leung, Lambert, and Mosenthal, 2004). It is proposed that risk assessment be conducted at both a system and asset-specific levels, since the former would help identify the critical assets in the network, while the latter can focus on analyzing a particular asset.

From the freight railroad network perspective, the system would include: nationwide railroad network (yards, tracks, sidings, etc.), hazmat commodity flow data, and the spatial location of population centers. The above information could be used to simulate various hazmat release scenarios and the associated consequence, and develop a risk–severity matrix that could be combined with "intelligent

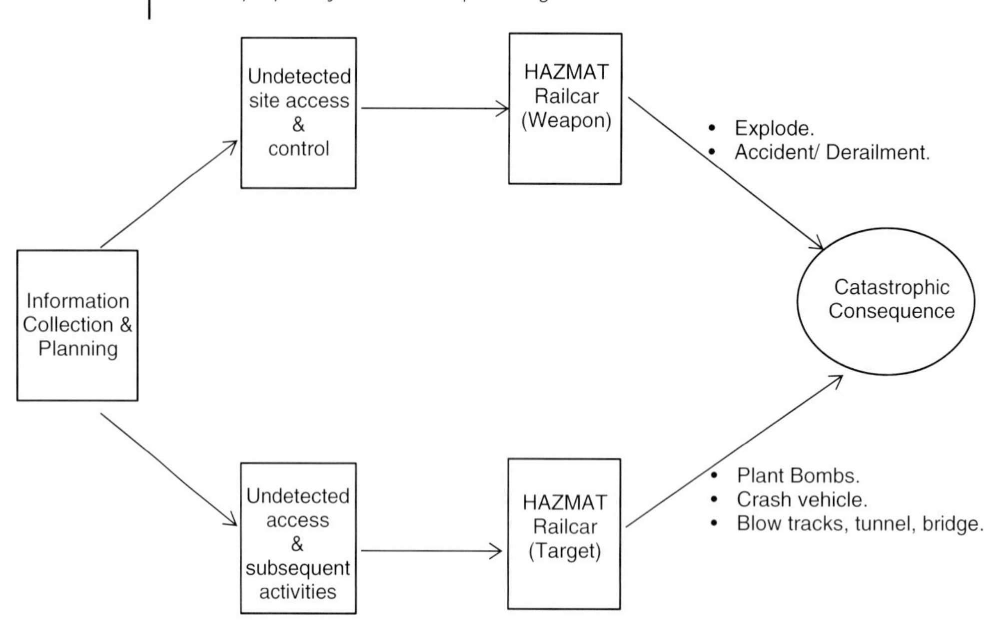

Figure 5.3 Possible events resulting in catastrophic consequence.

estimates on terrorist attack or intrusion probabilities" to ascertain the *critical* points in the system. These critical points would then be analyzed for vulnerabilities to attacks and/or failures. It is important to note that the proposed approach is not the only way to identify critical points, and that any methodology will pose significant challenges. This is because, in general, hazmat events are fairly rare – where expert opinion and Bayesian methods have been used with varying degrees of success, but the efficacy of such techniques to estimate "terrorist attack" is unproven.

For exposition purposes, Figure 5.3 depicts possible sequences of events resulting in catastrophic consequence. The depiction is motivated by the probabilistic risk analysis (PRA) method, initially developed for the purpose of assessing the safety of nuclear reactors, but adapted to study other situations in which catastrophic failures are possible (Harris, 2004). For the catastrophic consequence to perpetuate, each event in the sequence should fail for the law-enforcement agencies (or equivalently succeed for the terrorists). This also implies that catastrophe can be avoided if the terrorist (or intruder) is intercepted at one of the preceding stages. For example, a catastrophic consequence would be a concern if a terrorist manages to successfully collect the pertinent information and develop a plan to gain control of hazmat railcars (i.e. as a weapon), which could then be exploded. Alternatively, hazmat railcars could be targeted close to population centers by planting bombs, blowing tracks, or crashing vehicles such that maximum damage could be inflicted. However, the terrorist may be prevented from carrying out an attack if the security procedure succeeds at any stage before the last node. If the

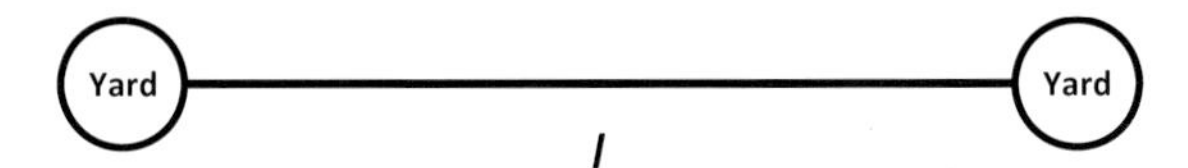

Figure 5.4 Rail track between two yards.

(conditional) probabilities that law-enforcement and intelligence agencies can gather the relevant information for each node can be evaluated, then the probability of failure for that specific node can be estimated. In the absence of appropriate data, estimating such probabilities is a challenging exercise for even a small network, let alone the vast railroad infrastructure in the United States and Canada.

We next outline a methodology that draws upon the low probability-high consequence events from hazmat logistics to assess "terrorist risk". Traditionally, hazmat transport risk is defined as the expected undesirable consequence of the shipment, that is, the probability of a release incident multiplied by its consequence (Erkut and Verter, 1998).

We adapt the traditional definition of hazmat transport risk to measure "terrorism risk" as follows: Consider a section of rail-track l between two rail-yards (Figure 5.4), where a hazmat railcar could be used as a weapon at either the yards or on the rail-track. At the yards, detonating devices could be used to maximize damage from a hazmat explosion. On the other hand, for rail-tracks it is not unreasonable to assume that a successful hijacking of a freight train with hazmat railcars is extremely difficult. This is because movement of freight trains beyond certain geographical limits requires gaining control of switches, and secondly the advanced communications devices in place at different links in the network would detect such a hijack and subsequently lock rail-tracks to prevent any movement. But the terrorist could still choose to target freight-trains carrying hazmat railcars by bombing the rail-tracks, crashing vehicles onto the freight trains, and/or derailing a train.

If the probability of an attack on rail-track section l is given by $P(TE_l)$, and the resulting consequence by $C(TE_l)$, then the terrorist risk posed by the hazmat railcar can be represented by:

$$\text{Risk}_l = P(TE_l) \times C(TE_l) \tag{5.1}$$

5.3.1 Probability of Attack

As indicated earlier (Figure 5.3), a successful terrorist attack on link l is only possible if each of the preceding layer of intelligence fails, which is schematically depicted in Figure 5.5, where: $P(IC')_l$ is the probability that the terrorists successfully collect information; $P(SA/IC')_l$ is the probability that intelligence agencies thwart the attempt to gain site access and control, while the complementary probability of success for terrorists is $P(SA'/IC')_l$ $P(I'/IC', SA')_l$ is the conditional probability when the terrorist could not be intercepted, and hence is in a position to inflict catastrophic consequences by attacking the rail-track (or the yard). Note that

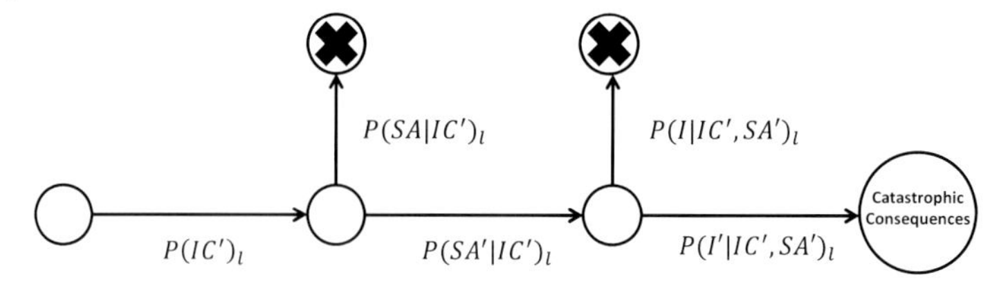

Figure 5.5 Failure sequence resulting in successful attack.

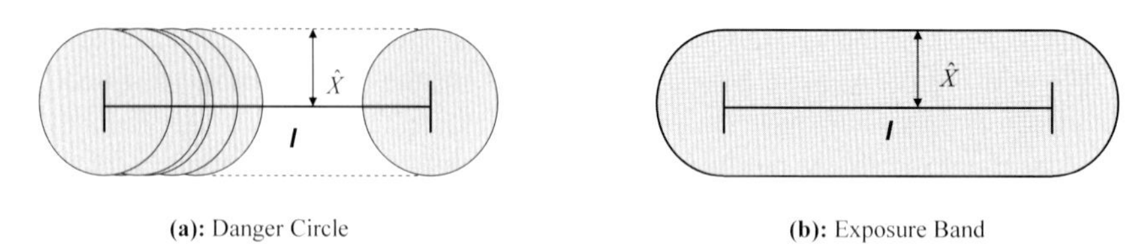

Figure 5.6 Impact area (a): danger circle; (b): exposure band.

a successful terrorist attack on link l would imply that the three preceding intelligence layers have failed to intercept the plot, and hence $P(I'/IC', SA')_l = P(TE_l) = P(IC')_l \times P(SA'/IC')_l$.

5.3.2 Consequence

Given our assumption that terrorist attacks are intentioned to inflict catastrophic consequences, one could make use of the work done in the context of hazmat logistics. For hazmat, consequences are a function of the impact area (or exposure zone) – and population, property, and environmental assets within the impact area. The shape and size of an impact area depends not only on the substance being transported but also on other factors, such as topology, weather, and wind speed and direction. Estimating, *a priori*, the impact area of a potential accident is difficult, and hence researchers have used different geometric shapes to model the impact area such as a band of fixed width around each link (e.g., Batta and Chiu, 1988), with a substance-dependent radius centered at the incident location (e.g., Erkut and Verter, 1998), and an ellipse shape based on the Gaussian Plume Model (e.g., Patel and Horowitz, 1990).

Perhaps the most common approximation of the impact area is the *danger circle*. By moving the danger circle along a route between two nodes, we get the fixed-bandwidth approximation (Figure 5.6). Since the bandwidth or radius is substance dependent, the consequence is limited to the area that needs to be evacuated or isolated. For example, according to the North American Emergency Handbook,[3]

3) North American Emergency Response Handbook, http://hazmat.dot.gov/pubs/erg/gydebook (accessed 20 December 2010).

800 m around a fire that involves a chlorine tank, railcar or tank-truck must be isolated and evacuated. Therefore, people within the predefined threshold distance from the railroad are exposed to the risk of evacuation (i.e. *population exposure*).

Note that since the objective of the terror attack is to inflict maximum damage, the representation of impact area as a danger circle – centered at the point of attack – is also appropriate from a security perspective. In addition, it is reasonable to assume that the exposure band is not appropriate since a terrorist would most likely target hazmat railcars close to major population centers, that is, a singular danger circle centered at the point likely to cause catastrophic consequence. In any event, estimating $C(TE_l)$ should take into consideration two things: *first*, the probability of targeting hazmat railcars in the freight train; and *secondly*, the presence of multiple railcars with hazmat (i.e. volume on-board).

5.3.2.1 **Probability of Targeting Hazmat Railcars**

Typically, a freight train carries a number of railcars, and not all of them derail and release as a result of an accident. Following a detailed analysis of FRA accident data, it was noticed that derailment probability is different for each position in a train-consist, and that it depends on the train length (Verma, 2011; Bagheri, Verma, and Verter, 2011). This is also pertinent from the security perspective, since the probability of targeting hazmat railcars in a train-consist will depend on their number and location along the length of the train. Now, assuming that a successful targeting of any hazmat railcar results in total loss of lading (or destruction), we have:

$$C(TE_l) = P(H/TE_l) \times \text{Impact}_l \quad (5.2)$$

where, $P(H/TE_l)$ is the probability of successfully targeting a hazmat railcar given the terrorist attack on rail-track l; and Impact_l is the impact on population and environment due to adverse impact from targeting hazmat railcars on rail-track l.

Before outlining the discussion on hazmat volume and the resulting consequence, it is important to note two things. *First*, one of the clear challenges would be to estimate conditional probabilities, since there is little or no empirical dataset, and hence expert judgements may be appropriate. *Secondly*, the impacts of most hazmat on human life and their interaction are unknown, which is relevant since railroads generally transport multiple hazmat, and proper risk assessment would require development of an approximation technique. Fortunately, we are interested in tracing the worst-case scenario and hence evaluations could be based on the hazmat likely to be the most detrimental. Such an approach will: preclude underestimation of risk; facilitate better emergency preparedness since the hazmat being considered is likely to cause maximum damage; and, offset the adverse impact of not keeping track of individual hazmat. We next outline the methodology needed to estimate the *impact on population* from multiple hazmat railcars.

5.3.2.2 **Hazmat Volume**

The use of a fixed-bandwidth approach, which implicitly assumes a standard hazmat volume, is inappropriate for estimating the number of people put at risk due to railroad shipments (Verma and Verter, 2007). The number and location of hazmat railcars vary considerably among trains, and hence it is important that the impact on population be a function of the volume of hazmat on-board. For example, a train carrying one chlorine tank car could potentially put at risk individuals within 1 km from the site of terrorist attack, whereas those further away may be at risk from successful targeting of more than one chlorine tank car. Note that successful targeting of hazmat railcars will result in an instantaneous explosion followed by a period where the lethal chemicals would drift into the atmosphere. While the casualty resulting from the explosion could be estimated by using representative scenarios developed for the chemical industry, the airborne feature could be captured through one of the air-dispersion models proposed for transport risk assessment of hazmat that becomes airborne on release. For example, a Lagrangian-integral dispersion model was used to model impact zones for six toxic-by-inhalation materials (Hwang *et al.*, 2001), a dense-gas dispersion model for comparing chlorine delivery option for rail and truck (Leeming and Saccomanno, 1994), and the Gaussian Plume Model (GPM) for risk assessment of railroad transportation (Verma and Verter, 2007).

Verma and Verter (2007) adapted the standard GPM to represent a single hazmat railcar (release source), which was then extended to model train shipments involving multiple hazmat railcars. It was assumed that the hazmat railcar and the impact point are at zero elevations, and hence the single railcar model was:

$$C(x, y) = \frac{Q}{\pi \mu \sigma_y \sigma_z} \exp\left(-\frac{1}{2}\left(\frac{y}{\sigma_y}\right)^2\right) \tag{5.3}$$

where, $C(x, y)$ is the contaminant level (parts per million) at impact point (x, y) in steady state; Q is the release rate of pollutant (mg/s); u is the average wind speed (m/s); σ_y the horizontal dispersion coefficient (m), $\sigma_y = ax^b$; σ_z the vertical dispersion coefficient (m), $\sigma_z = cx^d$; x is the downwind distance from the source (m); and, y the crosswind (perpendicular) distance from the source (m).

In estimating the steady-state contaminant level at point (x, y), the model assumes that the release rate and atmospheric conditions remain constant over the period of dispersion. Although the steady-state conditions are rarely reached, this is a common assumption – particularly reasonable during the first hour of release.[4),5)] It can be shown that, for any given distance, the contaminant level will be the highest at downwind locations (i.e. when $y = 0$). This implies that conservative risk assessment will simplify Eq. 5.3 to:

$$C(x) = \frac{Q}{\pi \mu \sigma_y \sigma_z} \tag{5.4}$$

4) Environment Protection Magazine, www.environmental-center.com (accessed 20 December 2010).

5) Environment Software and Services, http://www.ess.co.at/AIRWARE/gauss.html (accessed 20 December 2010).

Now, to extend the adapted GPM for a single railcar to that for multiple release sources the authors made use of earlier works in the literature. Pasquill and Smith (1983) and Arya (1999) suggested that contaminant from an array of sources with an arbitrary distribution of position and strength of emission can be modeled by superimposing the patterns of contaminant from these sources, and hence aggregating the resulting level at each impact point. In addition, they outline an approximation methodology to presume that all sources of release are positioned at the *hazmat-median* (i.e. the center of the hazmat block) of the train. After incorporating these two attributes, aggregate release from multiple sources can be determined by:

$$\bar{C}_n(x) = \frac{\sum_{i=1}^{n} n_i Q_i}{\pi \mu a c x^b x^d} \tag{5.5}$$

where $\bar{C}_n(x)$ is the aggregate concentrate level at downwind distance x due to hazmat released from n different railcars; and, n_i is the number of hazmat railcars with a release rate of Q_i. Other parameters are as before. If the immediately dangerous to life and health level (IDLH) of the hazmat being transported is known, then Eq. 5.5 can be rearranged to yield the expression for impacted area (i.e. radius of danger circle). For example, if $\tilde{C}$ is the IDLH of the hazmat being transported, then Eq. (5.5) will result in Eq. (5.6):

$$\tilde{X} = \sqrt[b+d]{\frac{\sum_{i=1}^{n} n_i Q_i}{\pi \mu a c \tilde{C}}} \tag{5.6}$$

Note that Eq. (5.6) aggregates contaminants from multiple sources of release, and thus captures the volume of hazmat released. The evacuation distance, $\tilde{X}$, determines the population centers likely to be exposed due to the terrorist attack on train-track l, which could be used to estimate the population impacted.

5.3.3 Route Risk

The accident rate and consequence modeling framework introduced in the preceding subsections can be extended to assess the terrorist risk for a complete route (or, path). For railroads, a route is a collection of rail-links and intermediate yards connecting the origin and destination for the specific freight train. For example, the route in Figure 5.7 is comprised of rail-tracks l and $l + 1$. In hazmat logistics, the travel on this route would be deemed as a probabilistic experiment, since the expected risk on rail-track $l + 1$ depends on whether the train met with an accident on link l. Hence, the expected consequence associated with this route can be determined by: $P(A_l)C(A_l) + [1 - P(A_l)]P(A_{l+1})C(A_{l+1})$, where, $P(A_{l+1})$ is the probability of meeting with an accident on link $l + 1$, and $C(A_{l+1})$ is the resulting consequence.

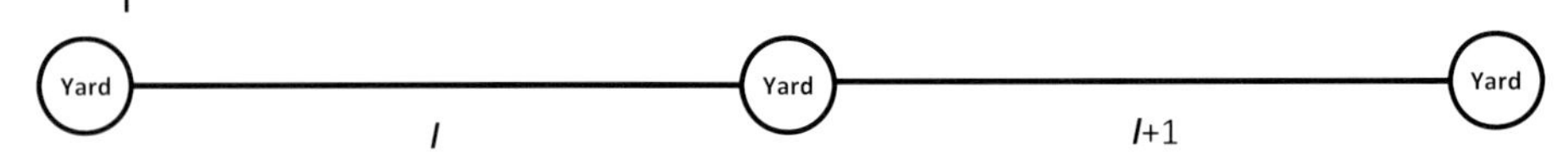

Figure 5.7 Freight service route.

Now, the security perspective entails a different objective, which in turn will impact the risk assessment approach. Note that the terrorist can attack rail-tracks l and/or $l+1$, but unlike hazmat logistics, the occurrence probability of the latter track section is not dependent on an unsuccessful attempt on the former. It should be clear that the probability of attacking any rail-track depends, among other things, on passing undetected through all the preceding layers, and the magnitude of expected damage. This is because for a terrorist, who has managed to successfully evade all security layers, the probability of successfully targeting any rail track is equal–although the intent to inflict horrendous consequences would be realized only if hazmat railcars are targeted close to dense population centers. This implies that rail tracks (or points on them) running close to or through population centers are most likely to attract attention of terrorists interested in catastrophic consequences (i.e. high probability of terrorist attack). In Figure 5.7, if $C(TE_{l+1}) > C(TE_l)$, then we should expect $P(TE_{l+1}) > P(TE_l)$, everything else being constant for the two rail-track sections. This is important since commensurate risk-mitigation strategies and/or emergency preparedness could be planned accordingly.

5.4 Risk Management

Because of the challenges associated with procuring good (and complete) intelligence data on terrorism, rail security remains an exercise in risk mitigation, as opposed to risk prevention. To that end, both the railroad industry and the government have pooled their complementary perspectives and capabilities to develop extensive mitigation strategies. We first outline the steps that have been taken as a result of the 9/11 events, and then delineate what else could be done to further mitigate risk.

5.4.1 Steps Taken

5.4.1.1 Information Sharing and Coordination

Following 9/11, the US administration was primarily focused on transportation modes considered to carry higher risk (Johnston, 2002), in particular air transport, which led to the creation of the Transportation Security Administration (Frederickson and LaPorte, 2002). The unregulated railroad industry took the initiative by putting together a team of experts, under the aegis of the Association of American

Railroads (AAR), expected to develop industry-wide strategies to assess risks and deal with potential threats in regard to hazmat, protection of the physical infrastructure, liaison with the military, operational security measures, and security of information technology and communications. The Information Sharing and Analysis Center (ISAC) was created to interface with government law enforcement, intelligence agencies, and computer emergency-response teams for threat and vulnerability information, which was shared with individual railroad operators for optimizing the use of their resources to protect the most likely targets of attacks (Hartong, Goel, and Wijesekera, 2008).

5.4.1.2 **Policing and Surveillance**

Railroads depend on their own police force (with full police and arrest authority on all railroad property), and train crews/roadway workers – who by virtue of operating over specific routes are intimately familiar with the territory and hence "normal behavior", to provide infrastructure security.

5.4.1.3 **Routing of Hazmat**

Railroads also carefully plan the movement of hazardous freight at all times, regardless of the security level. This is because movement of hazmat not only represents a potential for significant adverse consequences to the community and the environment, but also constitutes tremendous financial liabilities to the railroads in the event of accidental or deliberate release. To further mitigate the consequences of hazmat release, specific guidelines for routing hazmat and toxic-by-inhalation (TIH) materials have been developed (AAR Circular OT-55-I, 2005). Under this, member railroads are responsible for tracking the location of hazardous and TIH shipments from shipper to consignee, and ensuring the timely delivery of the material in accordance with the regulation. In addition, it establishes a mechanism for the railroads to provide, on request by a jurisdiction's public safety officials, a list of top 25 hazardous materials that are transported through the jurisdiction (Hartong, Goel, and Wijesekera, 2008).

5.4.2
Further Measures

5.4.2.1 **Interdiction Models**

As indicated earlier, the vast infrastructure of the railroad industry in the US and Canada render terrorist risk prevention a daunting task. Moreover, it is impractical from both time and resource standpoints to have layers of security and intelligence apparatus at each and every point in the network. One of the ways to get away from such a time- and resource-intensive approach is to make use of interdiction models developed in location theory to identify "critical points" in the network. Interdiction models with fortification are most useful in a setting involving intentional disruption (i.e. terror attacks), and can be used to identify the critical links or assets in the system (Scaparra and Church, 2008). A key question in fortification models is to identify which facilities to protect or fortify in order to preserve the functionality of the system as much as possible in the wake of external disruptions (Aksen, Piyade, and Aras, 2010). This is equivalent to deciding which rail-tracks and/or yards should have fool-proof safeguards to dispel any terrorist attack. Now, since these critical assets and/or links would be fortified using intelligence and surveillance, it is reasonable to assume that the terrorist may target less well-defended points in the infrastructure. Hence, it is important that terrorists are unable to predict even the less well-defended targets, which could be ensured by random assignments of security personnel and/or other protective measures. Predicting vulnerable targets and random assignment of security resources is equivalent to a nonzero sum game between two parties, but more importantly it is a very important area that has been unexplored, and requires further investigation.

5.4.2.2 **Tank-Car Design**

The last twenty years has witnessed a number of efforts geared towards risk mitigation with a focus on reducing the frequency of tank-car accidents and likelihood of release (Raj and Pritchard, 2000; Barkan, Treichel, and Widell, 2000). The two indicated studies, not constituting an exhaustive list, just like others have been conducted from hazmat logistics perspective, but do have relevance for "security". This is because detonating a bomb on a track can result in train derailments and/or collision with other trains. Consequently, insights gained from all these studies could indirectly improve security by augmenting the structural integrity of rail tank cars, which in turn would increase their resistance to compromise and subsequent material release (Hartong, Goel, and Wijesekera, 2008).

Placement of hazmat railcars has been investigated as a way to mitigate hazmat transport risk, but the related insights can indirectly improve security. Woodword (1989) made use of train speed and train length to conclude that separating hazmat railcars in a train decreases the probability of multiple railcars being derailed for small accidents involving relatively few railcars. This implies that adverse impact from bombing of rail-tracks could be minimized by ensuring maximum possible separation between hazmat railcars, since the probability of involving multiple railcars could be reduced. It has also been suggested that the front of the train is more prone to derailment under loaded conditions, and hence hazmat railcars should be placed near the rear of the train (Vole Transportation Systems Center, 1979; Battelle Columbus Division, 1992). Building on these works, Verma (2011) analyzed around 25 000 train accident records to conclude that 7th–9th train deciles are the safest position to place hazmat railcars for freight trains of any length. It is easy to see that since some terrorist attacks may result in train derailments, the insights obtained from the above studies could be useful to mitigate risk. Finally, regulators and railroad industry should consider developing concealment/masking techniques, so that the exact identity of hazmat railcars cannot be ascertained just by looking at the shape and/or placards. Of course, this would have to be done without undermining the original intent behind mandating placards.

5.4.2.3 **Shipment Routing**

The final measure to mitigating security (or safety) risk is by choosing the best route to move hazmat and regular shipments on the given railroad freight network. Glickman (1983) was perhaps the first to suggest that railroads should take into account population exposure when making operating decisions about routing and scheduling trains with hazmat. It was shown that rerouting, with or without track upgrades, is a feasible but potentially costly way of improving hazmat safety. While a dataset containing approximate flow patterns of hazmat was generated, it was assumed that all hazmat are equally hazardous and only residential population exposure was considered. In a more recent work (Glickman, Erkut, and Zschocke, 2007), transportation-network modeling has been combined with risk assessment to examine a sample of intercity rail routes to conclude that potentially safer routes at reasonable cost do exist.

Clearly, both the security and safety perspectives of hazmat transportation problems would be multiobjective in nature, and hence ideally a set of nondominated solutions (or *Pareto-optimal*) should be determined. A Pareto-optimal solution is one where one cannot improve on one objective without worsening at least one other objective. Although routing and scheduling of regular freight is well studied in the railroad transportation domain (Cordeau, Toth, and Vigo, 1998), multiobjective routing of hazmat shipments is not. To the best of our knowledge, Verma, Verter, and Gendreau (2011) and Verma (2009) are the only refereed publications dealing with risk and cost objectives to route hazmat shipments, and form the basis of the discussion in this section.

Verma, Verter, and Gendreau (2011) tackle a tactical planning problem faced by a railroad company that regularly transports a predetermined amount of regular and hazmat freight across a railroad network (Figure 5.2). So the objective is to determine the number and makeup of each type of train service, and the itineraries for each shipment such that the transport cost and the transport risk are minimized for the given set of demand. While the transport cost addresses the concern of the railroad company, population exposure was proposed as a measure of transport risk to capture the perspective of the other stakeholder, that is, regulatory agencies. The risk-assessment framework developed in (Verma and Verter, 2007), which estimated exposure zone as a function of hazmat volume on the train, was used to determine transport risk. Since the proposed optimization program lacked a closed-form expression for the risk objective, it was suggested that a metaheuristic solution methodology be developed to solve realistic size problem instances. Consequently, a memetic algorithm (MA), that combined both global and local searches, was developed. Subsequently, the proposed framework was used to solve realistic size problem instances based in Midwest United States, and also to develop a portion of the (quasi) Pareto frontier that could be used by make judicious decisions.

Although Verma, Verter, and Gendreau, (2011) and Verma (2009) propose analytical approaches to route hazmat shipments using transport risk and transport cost as the two criteria, both visited the issue from the hazmat-safety perspective. Now, in an effort to incorporate the specific attributes of "security", we make an attempt to adapt the framework presented in (Verma, 2009) to outline an optimization program that approximately captures the security dimension. To that end, we introduce the relevant notations, next.

Sets and Indices:

M: set of hazmat to be moved in the network, indexed by m.
L: set of origin yards, indexed by i.
J: set of destination yards, indexed by j.
C: set of classification yards in the network, indexed by c.
T: set of transfer yards in the network, indexed by t.
Z: set of train services in the network, indexed by z.
K_{ij}: set of itineraries connecting yards i and j, and indexed by k.
K_z: set of itineraries using train service z.
K_c: set of itineraries using classification yard c.
K_t: set of itineraries using transfer yard t.

Decision Variables:

$H_{ij}^{k,m}$: number of railcars with hazmat m using itinerary k to travel between yard i–j.
R_{ij}^{k}: number of railcars with regular freight using itinerary k to travel between yard i–j.
N^z: number of trains of type z needed in the network.

Parameters:

$\text{Risk}_{ij}^{k,m}$: security risk due to a railcar with hazmat m using itinerary k to travel between yard i–j.
$C_{ij}^{k,m}$: cost of moving a railcar with hazmat m using itinerary k between yard i–j.
C_{ij}^{k}: cost of moving a railcar with regular freight using itinerary k between yard i–j.
FC^z: fixed cost to operate train service of type z.
h_{ij}^{m}: number of railcars with hazmat m demanded at yard j from yard i.
r_{ij}: number of railcars with regular freight demanded at yard j from yard i.
U^z: capacity of train service of type z.
U_c: capacity of classification yard c.
U_t: capacity of transfer yard t.

(TP)

Minimize

$$\sum_{m,i,j,k} \text{Risk}_{ij}^{k,m} H_{ij}^{k,m}$$

$$\sum_{m,i,j,k} C_{ij}^{k,m} H_{ij}^{k,m} + \sum_{i,j,k} C_{ij}^{k} R_{ij}^{k} + \sum_{z} \text{FC}^z N^z \tag{5.7}$$

s.t.:

$$\sum_k H_{ij}^{k,m} = h_{ij}^m \quad \forall i, j, m \tag{5.8}$$

$$\sum_k R_{ij}^k = r_{ij} \quad \forall i, j \tag{5.9}$$

$$\sum_{i,j,k \in k_{ij} \cap k_z} \left(\sum_m H_{ij}^{k,m} + R_{ij}^k \right) \leq U^z N^z \quad \forall z \tag{5.10}$$

$$\sum_{i,j,k \in k_{ij} \cap k_c} \left(\sum_m H_{ij}^{k,m} + R_{ij}^k \right) \leq U_c \quad \forall c \tag{5.11}$$

$$\sum_{i,j,k \in k_{ij} \cap k_t} \left(\sum_m H_{ij}^{k,m} + R_{ij}^k \right) \leq U_t \quad \forall t \tag{5.12}$$

$$H_{ij}^{k,m} \geq 0 \quad \text{integer} \tag{5.13}$$

$$R_{ij}^k \geq 0 \quad \text{integer} \tag{5.14}$$

$$N^z \geq 0 \quad \text{integer} \tag{5.15}$$

The first objective in Eq. (5.7) contains the security risk stemming from routing hazmat railcars. The second objective involves cost to transport railcars, and the fixed cost to operate each type of train service. Constraints (5.8) and (5.9) ensure that demand for both hazardous and regular freight are met. Constraint (5.10) states that the frequency of each train type is determined by the number of railcars, moving on different itineraries between the origin and destination yards, but using that particular train service. Constraints (5.11) and (5.12) specify that classification and transfer operations at yards cannot exceed the corresponding capacities. Finally, constraints (5.13)–(5.15) specify sign restriction on the variables.

While transport cost can be computed using the framework presented in (Verma, 2009), the assessment of terrorist risk will draw upon the technique outlined in subsections 3.1.1 to 3.1.3. More specifically, consider that each service-leg of any freight train consists of a number of track segments between a pair of yards, and that a railcar uses a sequence of train services and yard operations to move from the origin to the destination. Since a terrorist would like to inflict maximum damage, which essentially translates into targeting a hazmat railcar on rail-tracks and/or in yards very close to population centers. Such scenarios could be simulated and then the approximate "impact on population" computed for all possible target locations between a pair of yards, and the maximum of all the computed numbers could be the value of the security risk for that specific segment/route. This is equivalent to making some track section and/or yards in the freight network very risky, thereby forcing hazmat railcars to use alternate routes – as much as possible. Since the railroad network is relatively sparse there could be instances wherein avoiding the high-risk path may not be feasible, and such instances would necessitate fortifying critical links with adequate surveillance, and/or having commensurate emergency-response systems in place.

Alternatively, one could consider delivering the hazmat shipments to a nearby rail-yard, and then making use of trucking to move them to the final destination. In fact, given the possibility of catastrophic consequence, some major cities such as Baltimore and Washington are considering banning railroads from carrying hazmat within their jurisdiction. Such measures would force railroad operators to take a much longer route to move hazmat shipments, and the resulting cost increase would be passed on to the end customers. Given this situation, it is extremely important that appropriate trade-off analysis between terrorist risk and network cost be performed to both identify the set of nondominated solutions, and also to gain insights into the efficacy of spending more money and the resulting reduction in terrorist risk.

Clearly the treatment of hazmat railcar routing from the security perspective is extremely rudimentary, and much of the analytical exposition is inspired by the methodology developed for safety of hazmat shipments, which itself is rather recent. We believe that the intent here was to throw light on the level of treatment with the objective to ignite interest in this significant area, and also postulate that successful endeavors would require an interdisciplinary approach. Insights resulting from academic and industry engagement could be extremely useful for the "advanced risk-based routing model" project being jointly sponsored by the Federal Railroad Administration and the Association of American Railroads.

5.5 Conclusion

The railroad industry – crucial to the United States and Canadian economy – has long considered itself to be the safest and most secure mode of transportation, as well as the most efficient. In the United States railroads have carried hazmat for at least the past century, and the current estimate of the number of carloads is 1.7–1.8 million. Safety, not security, had been the guiding policy of the federal program and the railroad industry pre-9/11, but the events of 9/11 have added another dimension, that is, each hazmat railcar (or train) can potentially be a target or a weapon.

Although the rail industry and the government have undertaken extensive efforts since, protecting the railroad infrastructure would be daunting, because an absence of enough empirical evidence precludes a thorough understanding of the terrorist mindset. Fortunately, such low-probability–high-consequence events are integral to hazmat logistics, an active research area, which could be drawn upon. Admittedly, risk assessment from a security perspective is much more complex since it involves estimating (conditional) probabilities of a number of intangible events such as good and complete intelligence, requires cooperation of the railroad industry and governmental agencies, and at least some understanding of terrorist mindsets. It is important that critical locations in the network are identified, and appropriate countermeasures put in place to limit the adverse impact of any attack. In addition, security resources should be randomly assigned to less well-defended targets such that their presence cannot be predicted by the terrorists.

To sum up, successful protection of the rail infrastructure will require a multi-disciplinary effort appropriately complemented by the cooperation of the railroad industry and the governmental agencies. While the probabilistic risk analysis developed for catastrophic failures of nuclear power plants can be a good starting point, the security domain does present potential for applying a game-theoretic approach to analyzing the behavior and possible moves of terror outfits. In general, security aspects of railroad transportation of hazmat a rather unexplored area, which contains a number of interesting problems of varied complexity, and does deserve the attention of individuals from a variety of disciplines.

Acknowledgment

This research has been supported in part by two grants from the National Sciences and Engineering Research Council of Canada (Grants #312936 and #183631). Both authors are members of the Interuniversity Research Center on Enterprise, Network Logistics and Transportation (CIRRELT) and acknowledge the infrastructure provided by the Center.

Bibliography

AAR Circular OT-55-I (2005) Recommended rail operating practices of transportation of hazardous materials. Technical Report.

Aksen, D., Piyade, N., and Aras, N. (2010) The budget constrained r-interdiction median problem with capacity expansion. *Central European Journal of Operational Research*, **18**, 269–291.

Arya, S.P. (1999) *Air Pollution Meteorology and Dispersion*, Oxford University Press, Cambridge.

Bagheri, M., Verma, M., and Verter, V. (2011) An expected risk model for rail transport of hazardous materials, in *Risk Prevention for Environment and Human Society against Dangerous Goods Transport Accidents and Malicious Intents: Methods and Tools* (eds E. Garbolino, M. Tkiouat, N. Yankevich, and D. Lachtar), Springer. Forthcoming.

Barkan, C.P.L., Treichel, T.T., and Widell, G.W. (2000) Reducing hazardous materials releases from railroad tank car safety vents. *Transportation Research Record*, **1707**, 27–34.

Barnhart, C., Jin, H., and Vance, P.H. (2000) Railroad blocking: a network design application. *Operations Research*, **48** (4), 603–614.

Batta, R. and Chiu, S.S. (1988) Optimal obnoxious paths on a network: transportation of hazardous materials. *Operations Research*, **36** (1), 84–92.

Battelle Columbus Division (1992) Hazardous materials car placements in a train consist: prepared by Thompson, R.E., Zamjec, E.R., Ahlbeck, D.R. Technical report.

Citizen's for Rail Safety (2007) *Securing and Protecting America's Railroad System: U.S. Railroad and Opportunities for Terrorist Threats: Prepared by Plant, J.F., Young, R.R*, The Pennsylvania State University, Harrisburg.

Cordeau, J.-F., Toth, P., and Vigo, D. (1998) A survey of optimization models for train routing and scheduling. *Transportation Science*, **32** (4), 384–404.

Erkut, E. and Verter, V. (1998) Modeling of transport risk for hazardous materials. *Operations Research*, **46** (5), 625–642.

Federal Railroad Administration (2008) Office of Safety, http://safetydata.fra.dot.gov/officesafety (accessed 20 December 2010).

Frederickson, H.G. and LaPorte, T.R. (2002) Airport security, high reliability, and the problem of rationality. *Public Administration Review*, **62** (special issue), 33–43.

Glickman, T.G. (1983) Rerouting railroad shipments of hazardous materials to avoid populated areas. *Accident Analysis & Prevention*, **15** (5), 329–335.

Glickman, T.G., Erkut, E., and Zschocke, M.S. (2007) The cost and risk impacts of rerouting railroad shipments of hazardous materials. *Accident Analysis & Prevention*, **39**, 1015–1025.

Harris, B. (2004) Mathematical methods in combating terrorism. *Risk Analysis*, **24** (4), 985–988.

Hartong, M., Goel, R., and Wijesekera, D. (2008) Security and the US rail infrastructure. *International Journal of Critical Infrastructure Protection*, **I**, 15–28.

Hwang, S.T., Brown, D.F., O'Steen, J.K., Policastro, A.J., and Dunn, W. (2001) Risk assessment for national transportation of selected hazardous materials. *Transportation Research Record*, **1763**, 114–124.

Johnston, V.R. (2002) Air transportation policy and administration at the millennium. *Review of Policy Research*, **19** (2), 109–127.

Leeming, D.G. and Saccomanno, F.F. (1994) Use of quantified risk assessment in evaluating the risks of transporting chlorine by road and rail. *Transportation Research Record*, **1430**, 27–35.

Leung, M., Lambert, J.H., and Mosenthal, A. (2004) A risk-based approach to setting priorities in protecting bridges against terrorist attacks. *Risk Analysis*, **24** (4), 963–984.

Milazzo, M.F., Ancione, G., Lisi, R., Vianello, C., and Maschio, G. (2009) Risk management of terrorist attacks in the transport of hazardous materials using dynamic geoevents. *Journal of Loss Prevention in the Process Industries*, **22**, 625–633.

Oggero, A., Darbra, R.M., Munoz, M., Planas, E., and Casal, J. (2006) A survey of accidents occurring during the transport of hazardous substances by road and rail. *Journal of Hazardous Materials*, **133A**, 1–7.

Pasquill, F. and Smith, F.B. (1983) *Atmospheric Diffusion*, 3rd edn Ellis Horwood, Chichester.

Patel, M.H. and Horowitz, A.J. (1990) Optimal routing of hazardous materials considering risk of spill. *Transportation Research Record*, **28A** (2), 119–132.

Plant, J.F. (2004) Terrorism and the railroads: redefining security in the wake of 9/11. *Review of Policy Research*, **21** (3), 293–305.

Raj, P.K. and Pritchard, E.W. (2000) Hazardous materials transportation on U.S. railroads. *Transportation Research Record*, **1707**, 22–26.

Scaparra, M.P. and Church, R.L. (2008) An exact solution approach for the interdiction median problem with fortification. *Computers & Operations Research*, **189**, 76–92.

Transportation Safety Board of Canada (2004) Statistical Summary Railway Occurrences, http://www.tsb.gc.ca/en/stats/rail/2004/statssummaryrail04.pdf (accessed 15 December 2010).

Verma, M. (2009) A cost and expected consequence approach to planning and managing railroad transportation of hazardous materials. *Transportation Research Part D: Transport and Environment*, **14** (5), 300–308.

Verma, M. (2011) Railroad transportation of dangerous goods: a conditional exposure approach to minimize transport risk. *Transportation Research Part C*, **19** (5), 790–802.

Verma, M. and Verter, V. (2007) Railroad transportation of dangerous goods: population exposure to airborne toxins. *Computers & Operations Research*, **34**, 1287–1303.

Verma, M., Verter, V., and Gendreau, M. (2011) A tactical planning model for railroad transportation of dangerous goods. *Transportation Science*, **45** (2), 163–174.

Vole Transportation Systems Center (1979) Strategic positioning of railroad cars to reduce their risk of derailment: prepared by Fang, P., Reed, H.D. Technical report: DOTITSC 7, 67.

Woodword, J.L. (1989) Does separation of hazmat cars in a railroad train improve safety from derailments? *Journal of Loss Prevention in Process Industries*, **2**, 176–178.

6
Security of Hazmat Transports by Inland Waterways

Pero Vidan and Josip Kasum

6.1
Introduction

Inland waterways are considered to be all navigation-passable waterways on rivers, lakes, canals that are arranged, marked and open for safe navigation. Inland waterways consist of segments of free flow without dams, segments managed by the dam system and canals and segments that are included in navigable parts of lakes. Hazardous materials started to be carried by inland waterways during the last two decades. Inland waterways extend deeply into the continent and sometimes pass through large cities and industrial centers. They are intersected by infrastructure potentially exposed to dangers such as:

- dams;
- bridges;
- hazardous material port terminals;
- hydroelectric power plants;
- nuclear plants; and
- other structures.

Inland waterways often serve for irrigation of agricultural land, cooling of nuclear plants, fishing, fish farming and so on. Apart from infrastructure, add-on risks are introduced by the transport and the storage of hazardous materials. Potential hazards may lead to endangering human health and loss of life, destruction of farms and industry, traffic suspension, floods, nuclear disasters, etc. The technology of safety and security of inland waterways is developing quite slowly as compared to the traffic increase[1]. Thus, we can assume that at present, the consequences of accidents, when they happen, would have a significant impact on the environment and economy of the exposed countries.

Security of inland waterways can be considered through the transport of hazardous goods and incidents in such areas. Transportation of dangerous goods is regu-

1) http://easyweb.easynet.co.uk/jim.shead/Inland-Waterways-of-England.html (accessed 11 November 2010).

Security Aspects of Uni- and Multimodal Hazmat Transportation Systems, First Edition. Edited by Genserik L.L. Reniers, Luca Zamparini.

lated by international treaties and conventions. Adequate protection of ports and inland waterways requires a new approach, as well as the implementation of new measures and technologies. It is therefore necessary to revise existing legislation and to invest in specific technology for controlling ports and waterways.

In this chapter, security improvements are suggested based on the use of new technologies onboard vessels and in ports, and the introduction of an "ISPSIW Code" is proposed, with the aim of introducing new measures of vessel security in international travel and ports that are open for international traffic.

6.2 Transport of Hazardous Materials by Inland Waterways – Current Legislation

According to European law regulations, hazardous materials on inland waterways are considered to be any materials classified under The European Agreement concerning the International Carriage of Dangerous Goods by Inland Waterways-ADN. For carriage and handling of dangerous goods, (either packaged, liquid, bulk, or liquefied gas) onboard and in ports, the following documents need to be complied with:

- the European Agreement concerning the International Carriage of Dangerous Goods by Inland Waterways-and;
- rules of technical supervision of the inland waterway vessels, ship registry;
- other national regulations.

The European Agreement concerning the International Carriage of Dangerous Goods by Inland Waterways-ADN has been adopted by the Central Commission for Navigation on the Rhine-CCNR) (CRUP, 2006).

Besides the ADN, other conventions that are applicable to maritime traffic and that refer to the carriage of dangerous goods have been the stronghold of development of regulation and provisions for inland waterway traffic. They are the following: International Convention of Safety of Life at Sea – SOLAS, International Convention for the Prevention of Pollution From Ships – MARPOL, Standards for Training and Watchkeeping Certificate – STCW, International Convention of Load Lines, International Regulations for Preventing Collision at Sea – COLREG, International Convention on Tonnage Measurement of Ships, International Aeronautical and Maritime Search and Rescue – IAMSAR and others[2]).

The International Maritime Dangerous Goods Code (IMDG) has been accepted as an international guideline for safe carriage or shipment of dangerous goods onboard. The aim of this regulation is the safety of the crew and the prevention of sea pollution. Regulations according to the 73/78 SOLAS and MARPOL conventions are obligatory for Member States of the United Nations. The regulations comprise the advice on terminology, packaging, marking, stowing, separation, handling, and on urgent interventions. As an example, a Dutch environmental consultancy indicates that safety equipment is often placed at unacceptable locations without risk analyses (Slob, 1998).

2) http://www.unescap.org/ttdw/Publications (accessed 20 February 2011).

6.3 Incidents on Inland Waterways

Safety of navigation can be seen as the measure of success of any sailing endeavor. The navigation should be carried out without endangering the lives of people, the vessel, cargo and environment. Incident measures are taken on the base of statistical data related to damages that occurred in particular areas.

The damages are usually expressed as a function of the losses of human life, the number of injuries, and also in terms of quantities and values of the damaged or lost cargo or vessel, etc. The safety of a vessel in navigation can thus be presented as a function of the form:

$$S_{bp} = f\left(b, S_{ec}, pp, \rho_p, \text{ost}\right) \tag{6.1}$$

in which:

b – safety of vessel and cargo
pp – safety of navigable waterways
S_{ec} – security of vessel and waterway
ρ_p – traffic density and
ost – other factors.

The incidents on inland waterways can occur as a result of:

- human negligence;
- *vis majeur*;
- willingly caused accidents.

Human negligence is the most frequent reason for accidents on inland waterways. They occur due to improper handling of technical aids, fatigue, lack of knowledge and negligence. In maritime traffic, human negligence has partly decreased due to the use of Safety Management Systems (SMS) and International Safety Management (ISM).

Vis majeur occurs as a cause for damage that cannot be influenced directly. These are for example, damages caused by hydro-meteorological phenomena such as floods, bad weather, change of water level, icing etc.

Willingly caused incidents do rarely occur but they are considered to be most dangerous because of the potential size of resulting damages. A willingly caused incident mostly happens due to:

- war;
- terrorism;
- vandalism;
- pilferage.

Modern inland waterway vessels intended for the carriage of dangerous cargoes are potential ecological bombs. They could cause extensive damage if the cargo were to get lost. Since the inland waterways are limited by land, they need longer times to establish the ecological balance. Rivers often have relatively strong water current and if an accident spreads rapidly, the accident's environmental consequences could affect a large area without appropriate response measures. Lakes (contrary to rivers) have no strong water current, and hence an accident's consequences do not have a tendency to spread faster. Nonetheless, due to poor water changes in such a lake ecological system, the ecological balance is difficult as well.

According to American Waterway Operators (AWO), a large increase of inland waterway traffic is expected. Waterway vessels could pose an important danger in the case of terrorism alert, especially due to the fact that they are not protected and that they sail through large cities.[3]

Potential pollution contaminates relatively large areas (Slob, 1998). Therefore, the recovery and stopping of pollution is more difficult. Potential prevention of oil spillage due to terrorism is difficult to predict and apart from the effect of generating casualties and material damage, there is also an effect of shock.

Although the motivation varies in acts of vandalism and pilferage, the types of damage they cause and the consequences for the environment are similar. The most frequent type of vandalism is deliberate valve opening on dangerous cargo tanks with the aim of releasing the hazardous material and causing environmental damage (Slob, 1998).

Pilferage is an attempt to take possession of obnoxious cargo with the possibility of pollution. For example, in Russia, several accidents have occurred due to pilferage with the consequences of environment pollution. The attempt of pilferage of an oil pipeline in southern Russia resulted in the spillage of 300 tonnes of crude oil into the river Giaga (Reuter Newswire, 1996). A similar incident happened in Ukraine in 1993 when the pilferage of oil resulted in the spillage of 72 tonnes of diesel fuel into the Uzh, the tributary of the river Tisza (Reuter Newswire, 1996).

6.4 Security of Inland Waterways and Ports – Current Practices

In endeavors to increase safety and security on inland waterways, modernization of inland waterways and river fleets has been noticeable. Europe pursues improvement and development of navigation in inland waterways by various development projects. The projects that could be pointed out are the project of development and establishment of a River Information System (RIS), supervision of navigation service – Vessel Tracing System (VTS), Automatic Identification System (AIS), and Inland Electronic Charts Display Information System (Inland ECDIS) and the like. An RIS is being established for all navigation areas of European Union inland waterways. Proper informing of vessels would greatly aid the increase of security on inland waterways.

3) For example, traffic was blocked following the NATO bombardment in Serbia in 1999 during which several bridges on the Danube were destroyed (Reuter, 2001, accessed 1 February 2010 at http://www.emperors-clothes.com/articles/rozoff/danube2.htm).

The security of vessels in inland waterways could be considered satisfactory if:

- the waterways have a high level of security;
- the vessel and crew have a high level of security;
- the cargo is not dangerous for transport; and
- other factors (such as: small distance to another vessel with inadequate level of security, the factors that constitute force majeure, the factors that are not covered by the research and could pose less impact on security) are favorable.

Navigable waterways security is considered satisfactory if there are means of information service such as RIS, VTS, Inland ECDIS, AIS, etc. Apart from the aforementioned statistical data of accidents in a particular period of time for satisfactory safe waterways, there should also be a service of obtaining data on relatively rare accidents and possible dangers.

The level of vessel and crew security is considered to be satisfactory if all the international provisions have been fulfilled, such as the provisions regarding vessel security, cargo, training, required publications and procedures. These provisions are included in the International Ship and Port Facility Security Code (ISPS). The code was adopted as an amendment of the 1974 SOLAS convention and it has been obligatory since 1st July 2004. It refers to war ships, government ships and government ships for noncommercial purposes. The ISPS code is obligatory on vessels in the international waterways, for passenger vessels, fast passenger vessels, cargo vessels, fast cargo vessels over 500 BT[4] and mobile off-shore platforms.

Measuring cargo dangerousness implies determining the levels of noxiousness for the crew's health, for the cargo, for the vessel and for the environment.[5] Other factors encompass unexpected and rare examples that are caused by *vis majeur*, which could endanger people, vessel and cargo. In ports and terminals one can come across various methods of handling of dangerous cargo. In spite of technological advancement regarding the security of people and the environment, ports and terminals still remain endangered in the case of attempts at diversion. The smallest disorder in the work of such facilities could cause the destruction of ports, the surrounding areas and cities. Using inland waterways for LNG transport is a particularly relevant option for landlocked countries in central Europe where river and canal networks are particularly dense, and where demographic or environmental specificities do not justify laying down pipelines. Another reason for resorting to LNG is the exploitation of small gas fields of limited capacity whose productive lifetime cannot justify building pipelines. Europe's inland waterways offer an attractive transport option for LNG. In fact, LNG deliveries by rivers and canals in containers could be seen as a virtual pipeline network in addition to traditional pipelines.[6] However, LNG goes hand-in-hand with substantial risks.

4) Brutto tons.
5) http://www.uscg.mil/../chapter1.htm (accessed 11 December 2009).
6) http://assembly.coe.int/Main.asp?link=/Documents/WorkingDocs/Doc10/EDOC12424.htm.

The fire around an LNG container, for example, could result in an explosion within the container, generated by an inner pressure build-up. This occurs when the integrated safety system would not be able to cope with the pressure build-up, and due to the heat created by the outside fire, the container is warming up to the point of disintegration. The result of that scenario is a massive explosion of the container, which is described as a boiling liquid expanding vapor explosion scenario (BLEVE).[7] As a result of the BLEVE, extreme damage to a ship's structures can be expected.

When a significant leak of LNG occurs from a cargo tank spreading a liquid (that is, LNG) on deck and further into the sea, no fire or explosion will result, but a large difference in temperature between the LNG and its environment will occur, creating a tremendously large amount of evaporation. Such a leak over the water in some cases can cause a rapid phase transition where LNG quickly transforms its state from liquid to vapor, causing a physical explosion.[8] Such an explosion can be demonstrated as a release of sound and pressure (without flame and heat), which can cause a certain amount of damage to the ship's/barge's hull and eventually to the surrounding infrastructure.

The measures for ensuring security of the vessels and ports have been determined by the Ship Security Plan (SSP). This security plan comprises the ship's drawing with marked critical security points, display of the areas limited to the access of unauthorized persons, display of the lights on the vessel, vessel's IMO designated number, data on the officer in charge of the company security, security plan and procedures, etc. For security reasons, the SSP content, which is highly delicate, is considered a secret and it is not within the scope of vessel inspection. The companies are, however, obliged to supply the vessel with obligatory equipment for the security of the vessel such as the following:

- AIS;
- security alarm system;
- lighting;
- means for limitation of the access to the vessel.

The AIS device transmits the vessel data by very high frequency (VHF) radio waves. The vessel data include the number, name, condition (when underway, at anchor, on berth, etc.), port of departure, port of destination, expected time of arrival into the destination port, number of crew, etc. The AIS device is operated by the vessel officer within his scope of work. AIS needs to be switched on at all times and the vessel data have to be updated. The source of feeding is the emergency 24 V network, intended for unhindered continuous operation.

The security alarm system is mandatory on all vessels on which the application of the ISPS code is obligatory. It has been designed as a silent alarm. It is transmitted from the vessel to the shore in a way that it does not alert the vessels in the vicinity. The position of the safety alarm system on board is secret, the position

7) http://www.science20.com/chatter_box/what_bleve-77206 (accessed 12 May 2010).
8) http://www.buergerimstaat.de/2_00/rhein.pdf (accessed 01.06.2010).

is drawn into SSP and it is known to the SSO and the master. In order to provide undisturbed operation, the source of electric energy used is 24 V.

The lighting system supplies the vessel with the lights as prescribed by SSP.

The means for limitation of the access to the certain locations on the vessel include locks, cameras, alarms etc. Metal and explosive detectors, protective voltage rail, etc., could be included in the optional equipment.

Harbor and port authorities are obliged to supply security of the vessel while in port. The SSO obligation is to announce the vessel's arrival at least 24 h before it arrives to the port, to make contact with port authority representative, to exchange the data on the security level, to submit the data on the vessel and crew, etc.

The SSO is obliged to announce the security level of the vessel in accordance with the security of the port. Depending on potential future events or expressed doubts regarding the danger, the level of vessel security could be changed. In that case, the SSO is obliged to inform the company and port officers in charge of security.

Also, it is obligatory to organize:

- access to the vessel;
- supervision of embarking of persons with luggage, checking of their identification papers and possible arms and dangerous goods that could endanger the security;
- supervision of the area with the access only to the authorized staff members;
- supervision of the areas on decks and surrounding the vessel;
- supervision of cargo handling and vessel holds;
- good functioning of communication regarding the security issues;
- drills according to the SSP;
- testing of equipment, security onboard and the like.

Although it is possible that in the near future inland waterway navigation becomes a target of possible terrorist attacks and that the consequences could be enormous, at present there have not been any concrete security measures taken for the security of inland waterway ports and vessels by the IMO, the International Waterway Organization (IWO) or any other relevant international organization.

6.5 Proposals for Security Improvements on Inland Waterways

Because of the many threats that occur on inland waterways the establishment is proposed of ISPS that would relate to the protection of inland waterways. The proposal for the working title is: "International Ship and Port Facility Security

Code in Inland Waterways" (abbreviated ISPSIW). When adopting the ISPSIW code for conventional and nonconventional inland waterway vessels, a development according to the division of basic levels is suggested:

1) conventional and nonconventional inland waterway vessels;
2) organization;
3) ports;
4) other (supervision and control systems).

The first level refers to conventional and nonconventional inland waterway vessels. Here, it is important to adhere to the basic principles of the ISPS Code, but with attention drawn to the specific problems of inland navigational waterways. The navigation in the obscure parts of rivers and lakes, passageways under bridges and the like, present an immediate danger for boarding of intruders and unwanted cargoes such as terrorists and explosive devices. The navy and public vessels may be considered sufficiently protected.

We may assume that the vessels for tourism, sport and leisure, rafts and fishing boats did not have sufficient navigational equipment and vessel security equipment. We also assume that the crew on these vessels have no proper training for navigation and control of these vessels. There are more such vessels than conventional ones and thus they are more difficult to be supervised (Kasum, Marusic, and Grzetic, 2006). There are differences in skills, required approvals and state control that vary from country to country. Therefore, harmonization of education programmes for crew training on vessels in inland waterways and nonconventional vessels, and international unification of the official form of various authorizations and supervising systems can be suggested.

In the second level of the ISPSIW Code development on the inland waterways, a need to identify various forms of organizations, legal forms of society and the like, exists. The inland waterway nonconventional vessels are mostly intended for sports and leisure. The owners of these vessels are often individuals or belong to the government. As the boats for sports and leisure which belong to individuals are numerally more, it is proposed to organize them into charter companies, sports clubs, diving clubs, etc. This makes it easier to control them both from companies and from foreign ports.

The third level of the ISPSIW Code development regards ports, ports of refuge and winter harbors. Ports of refuge and winter harbors are places adjusted for acceptance of vessels, receipt and shipment of cargo and passengers. Ports and winter harbors are usually constituents of larger industrial centers and cities. Because of their characteristics, they are protected from hydrographic and meteorological effects. River ports are usually ports in large cities, located near towns or in their centers. Therefore, the security aspect of such ports is important from the aspect of security of city population and safety of navigation. The access to ports from the land is not supervised or it is not sufficiently protected. Port approach is possible even with smaller boats without supervision. That is why the arrival of vessels into ports should be announced, and the time they will spend in ports, embarkment and disembarkment of people and cargo should constantly be monitored. The inland waterway ports should have the same security norms as the sea ports that are open for international traffic. The proposal is to equip the inland waterway ports with port radars, an AIS system, video supervision, a system for detecting container contents by indirect inspection, etc. It is recommended to train the teams for various types of security problems and risks. The Company Security Officer (CSO) is also proposed to be mandatory.

The fourth level regards the security of other important elements. They especially refer to various supervision and control systems. The supervision and control of river navigation is not performed in the same way everywhere. Some countries do not have complete records of vessel transits on their rivers. In some countries the river navigation is performed without announcement and without information on destination, type of voyage and nature of cargo. The country with jurisdiction on inland waterways should establish complete supervision of vessel navigation. For example, with establishment of coast and port radars, RIS should also be established.

Inland waterways affect a large number of industries in the field of metallurgy, oil derivatives, power plants, etc. The traffic in inland waterways is also constantly increasing. In inland waterway ports there are various infrastructure facilities such as oil terminals, terminals for liquid gas and chemical products, ores and various types of dangerous cargoes. The access to these ports is relatively easy, which increases the risk factor and potential danger. Therefore, we recommend developing and implementing a security system on inland waterways through international organizations such as the United Nations or IMO. However, the procedure of ISPSIW organization and implementation is very complicated.

ISPSIW is proposed through the development of the existing protective measures and mentioned levels (Figure 6.1).

The algorithm from Figure 6.1 starts with security of conventional and nonconventional ships. It is assumed that the inland conventional ships are protected by special measures that are prescribed by ISPSIW. For the security of nonconventional ships, it is necessary to develop specific measures that would be able to monitor these. In maritime transport, the ratio of the conventional and nonconventional ships reaches 1 : 1 200 000 (Kasum, Marusic, and Grzetic, 2006). Therefore, we recommend specific measures for the protection of these vessels (Kasum, Marusic, and Grzetic, 2006). The algorithm further proposes integration of these measures on all ports, companies and smaller boats (logistics, government, and shipping companies). A simulation of these measures under conditions of use of ISPSIW and their international applications for inland waterways is recommended.

To put it simply, an inland waterway can be regarded as a zone in the inland waters of certain width and depth in which navigation is carried out. Apart from navigational dangers such as fairway crossings, etc., navigable fairways of inland waterways are crossed by highway and railway bridges. They present additional hazards for navigation safety.

We propose using assessment measures of potential human-related risks, material-related risks and other risks that are not covered by the survey but could have an impact on security. Therefore, the density of various facilities on inland waterways is measured based on:

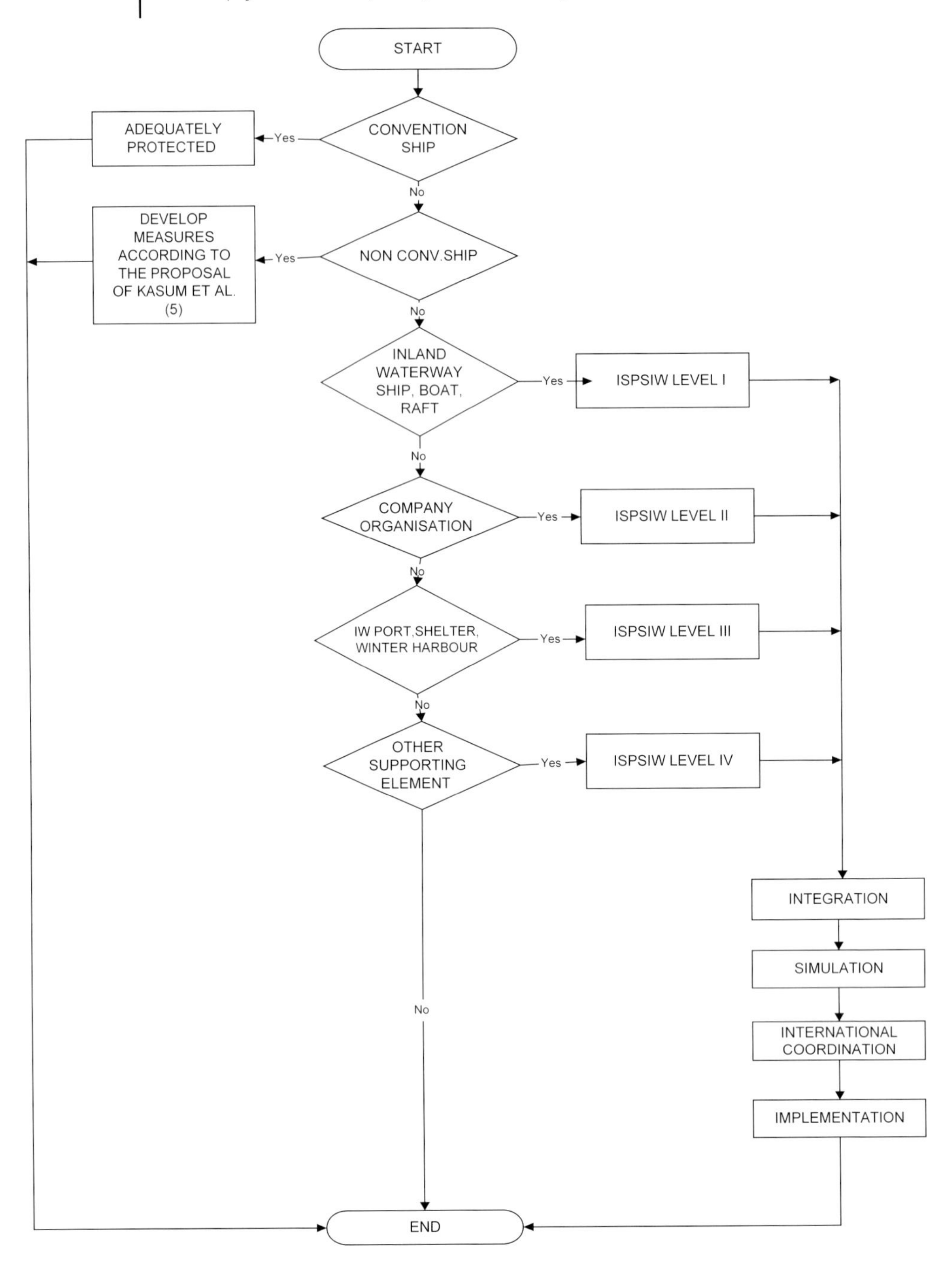

Figure 6.1 General ISPSIW development algorithm. Source: Kasum, Vidan, and Baljak (2010).

- density of populated places on a waterway G_{nm};
- density of population on a waterway G_{na};
- density of potentially threatened objects on a waterway G_{no}.

The density of populated places on waterways (G_{nm}) is defined as the ratio between the total number of populated places on the waterway (Nm) and the waterway length (l) as expressed by Eq. (6.2):

$$G_{nm} = \frac{\sum_{i=1}^{n} Nm_i}{l} \tag{6.2}$$

The density of population on a waterway (G_{na}) can be defined as the ratio between the number of inhabitants on the waterway (S) and the length of waterway (l) as expressed by the Eq. (6.3):

$$G_{na} = \frac{\sum_{i=1}^{n} S_i}{l} \tag{6.3}$$

The density of potentially threatened objects on a waterway (G_{no}) is considered equal to the ratio of the number of threatened objects (O) and the length of the waterway (l) as expressed by the Eq. (6.4):

$$G_{no} = \frac{\sum_{i=1}^{n} O_i}{l} \tag{6.4}$$

In order to increase security levels on inland navigable waterways, the results expressed as proposed measures could be included into the information content of river charts and navigational publications. For example, the proposed measure of the danger coefficient based on threatened objects could be denoted by:

- ○ mark;
- △ mark;
- □ mark.

The ○ mark designation denotes an area with a high level of security because of threatened objects, whereas the △ mark designation denotes an area with a medium level of security because of threatened objects, and the □ mark designation denotes an area with a low level of security because of threatened objects.

The examples of navigation in the Main and in the Rhine rivers show the assigned measure of danger coefficients (Figure 6.2). According to the available data,[9] it has been estimated that the industrial area of Frankfurt, the city of Frankfurt and the city access areas are considered to be dangerous for navigation because

9) http://www.buergerimstaat.de/2_00/rhein.pdf (accessed 01.06.2010).

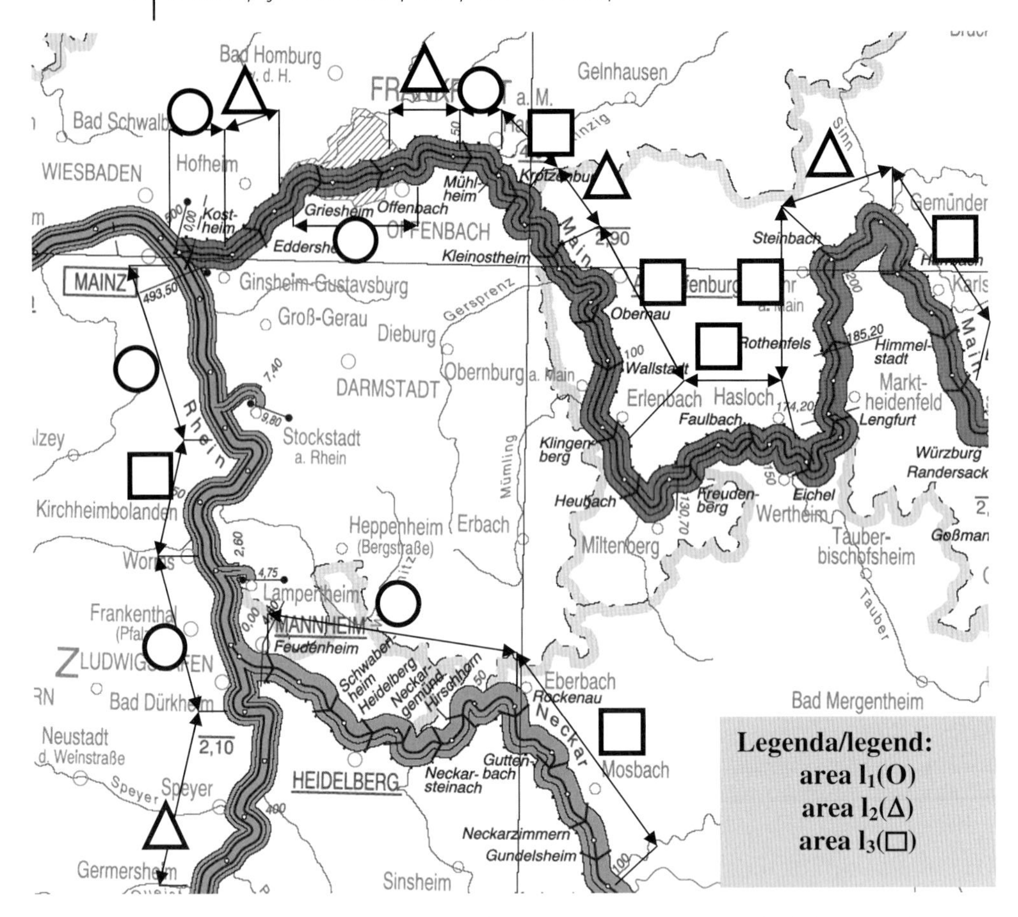

Figure 6.2 Example of area with different values of security danger levels because of threatened objects for navigable parts of the Rhine and the Main. Source: http://www.fgs.wsv.de/produkte/kartographie/argo_encs/index.html (accessed 1st April, 2000).

of threatened objects. The same applies to the bridges that are considered to be too low. It is therefore recommended for potentially dangerous segments of the river to limit the navigation to the daytime only with reduced speed and calling VTS, etc.

The coefficients are those of mean values in the immediate vicinity of the area with high security because of threatened objects. Such areas are considered to be the ones near the obstacles, such as locks, factories, bridges, canals and so on. In such areas it is obligatory to call VTS and to navigate with caution, etc.

The areas without dangers have relatively small values for danger by human error. Such areas are considered to be the ones of unobstructed navigation, the areas of medium and small traffic density and of safe depths.

Table 6.1 System of danger appraisal for safety and security of vessels in waterways.

Danger to navigation safety and security (S_g)	Navigation description	Appraisal
Great danger to navigation safety and security	Immediate or lasting navigation suspension due to obstruction of navigation.	3
Moderate danger to navigation safety and security	Navigation is possible only to the closest refuge or safe navigation waterway. Refuge or safe navigation waterway should be near to the current position. The navigation at this location should not include a navigation area appraised by level 3.	2
Small danger to navigation safety and security	Navigation that might be endangered by some of the factors. Navigation is possible with greatest caution.	1

Source: Vidan, 2010, Doctoral dissertations.

Hypothetical data for Figure 6.2 are the result of assessments. In order to obtain more detailed safety and security analyses, detailed hydrographic measurement would need to be carried out and estimations based on updated available navigation publications as well as security assessments by intelligence agencies should be collected.

The value interval of these coefficients can be determined by new hydrographic measuring of navigable waterways. Based on potential values it is necessary to determine the possible safety and security risk (S_g) of navigation waterways. The appraisals are presented in Table 6.1.

These appraisals should be included in nautical charts for particular areas. Based on the appraisal system, the security measures may be applied from ashore and from onboard. The measures based on the appraisal of safety and security danger may be implemented using the decision algorithm from Figure 6.3.

The algorithm in Figure 6.3 describes the inland waterways in varying degrees of safety and security. A S_g (degree of safety) is considered low (satisfied) if it is equal to level 1 (Table 6.1). In that case, there are no signs of safety and security danger for the ship, crew and cargo. For example, navigation on well-marked waterways, navigation on sufficiently wide and deep waterways, navigation in good weather and hydrological conditions.

S_g equals level 1 if there are no visible signs of danger, but if difficulties in the conduct of navigation, maneuvering and security can be expected. Such navigations can be passing near the port and anchorage, passing through the channel, navigation under bridges, and navigation close to industry, etc.

S_g equals level 2 if some safety and security dangerousness for the ship can be expected. Some examples of this level are passing through an area of intense traffic, the entrance to the harbor or shelter, passing through the locks, etc.

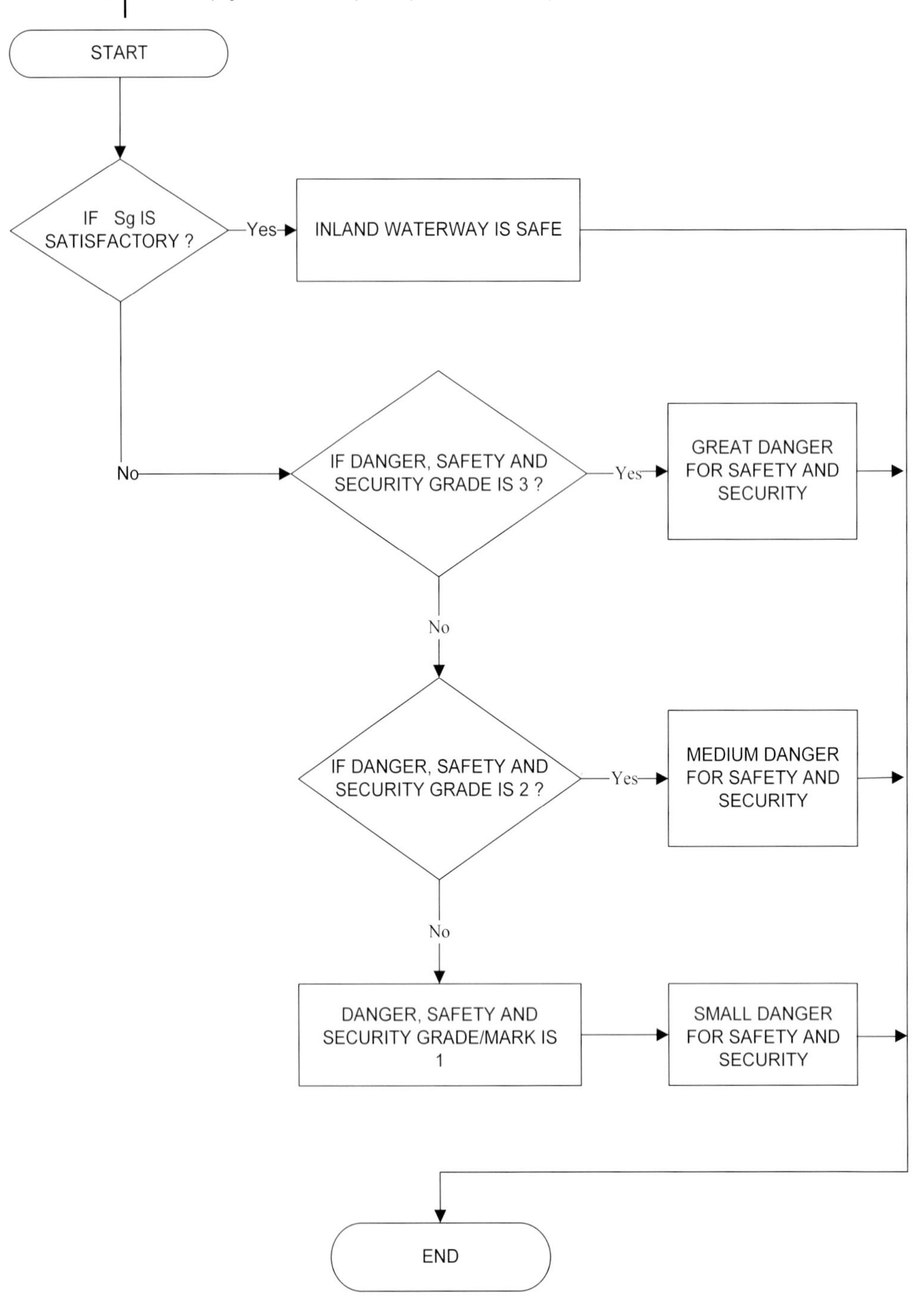

Figure 6.3 Safety and security decision and appraisal algorithm for inland waterways. (Source: Vidan, 2010, Doctoral dissertations).

S_g equals level 3 if safety and security of the ship is directly threatened or if it is expected that the ship's vulnerability could affect the environment. For example,: terrorist attacks, fires, explosions, collisions, leaking loads, and hijacking leads to such a level.

6.6 Proposals for Improvements of Inland Ports' Security

Many international sea ports are not ready to apply the ISPS Code. The tendency in sea traffic is the development of larger ports and the closing of smaller ones. Larger ports are easier to supervise than the smaller ones that are often dislocated. Every international port today is strictly supervised for possible emigrations/ immigrations of people, for attempts at contraband through the port, for attempts at pilferage of cargo and port inventory. Such trends are also expected for inland waterway ports. It is expected that with the development and application of the ISPSIW Code in inland waterways, the procedures in ports would become stricter.

Each port, subject to the ISPSIW code, should have an especially trained unit that could act in case of a terrorist attack (i.e. after security level 3 is announced). These intervention units should be familiar with explosive devices and they should use robots or armored crafts in order to deactivate explosive devices. In order to alleviate control of personal effects and smaller shipments that enter the port, the ports should be equipped with detectors for metal and explosives. All devices should be standardized, and IWO should prescribe the number of devices in accordance with the size of the port and number of arrivals and departures in ports. The vessels should be fenced with special wires with high voltage electricity. In ports and at anchorages that are considered to be endangered it is necessary to organize a special security service (see Figure 6.4).

As far as supervision is concerned, containers pose the greatest problems with loading and unloading cargo. Some seaports are already using the latest technical achievements during supervision of that cargo, thus the application of similar devices is proposed for inland ports as well (see Figures 6.5 and 6.6).

One of them is, for example, supervision by radiation in X-ray. One of the devices for detection of cargo content (metal, explosives, noxious cargo, etc.) is the so-called *Cargo Inspection System* as illustrated in Figure 6.5. This system can scan the cargo content in several radiographic pictures (shown in Figure 6.6). The time of scan duration for thirty 2 TEU containers or sixty 1 TEU container is 1 h and there is an option of fast scanning when all data goes directly into the computer memory, and is processed later. Such a system located at the exit from the port has the possibility of rapidly taking multiple radiographic pictures, and in combination with an optical sign reader, it can control the traffic and the contents of containerized goods passing through the port. The cheaper counterpart of such system is a scanning reader of empty containers. Apart from supervision of cargo in cargo ports, it is essential to improve the checking of passengers and vehicles in passenger ports. The ports in inland waterways are often insufficiently lit, and

Figure 6.4 Security wire against attacks used on dredgers on the Amazon river.

Figure 6.5 Mobile Cargo Inspection System. Source: ADANI, www.adani.by.

the access to the ports is not strictly supervised. Also, the supervision of a dangerous cargo entrance is not sufficient. Therefore, there should be built specially protected terminals that would have safety and security equipment and supervision. Such equipment would include infrared cameras, lighting, devices for supervision of passports and ship books, as are used in seaports and airports, metal and explosives detectors, etc.

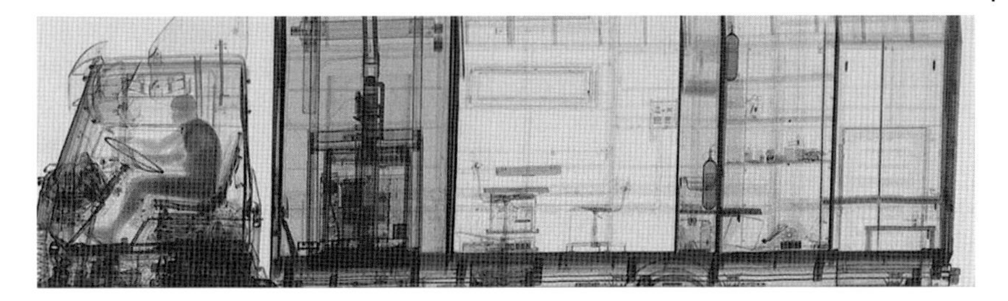

Figure 6.6 Display of container content. Source: ADANI, www.adani.by.

In that way, possible risks would be reduced for port terminals being usually part of a larger industrial plant. Terminals for transshipment of inflammable cargo (LNG; LPG; toxic materials; radioactive cargo, oil, chemicals, etc.) should be closed for night navigation. That would require the procurement of clearance inwards and outwards during the night. A similar measure has been used in the USA at LNG sea terminals.

6.7 Conclusion

There has been a trend of a decrease in the accidents in inland waterways, while an increase in traffic is being recorded at the same time. Nonetheless, it is justified to assume that it is essential to further investigate new methods, techniques and technologies in order to further increase safety and security. Although a reduction is expected in the safety-related likelihood of casualties, loss of ships and cargo, ecological incidents, etc., the possibility of a security attack towards ships becomes larger. Threats towards ships and inland waterways are real because they are considered by terrorists as poorly protected. An increase of traffic and the industry's dependence on this relatively cheap mode of transport make them even more vulnerable. The attacks to inland waterways could have a substantial impact on the economy of the region, and on the life of the local population. At present, dangerous goods, which include explosives, radioactive cargo, etc., can become a dangerous weapon if their transport is not properly protected. There are proposals for adjustment of these methods to the rivers, lakes and canal navigation, as well as for the new working title, ISPSIW rules.

The security in inland waterways has been determined by the state and international provisions. Their further development is recommended. Therefore, navigation security is described by security measures for particular areas of navigation waterways. The measures are expressed by coefficients that give an insight into traffic density, settlements, potentially endangered objects and being endangered by human error. By these coefficients it is possible to assess certain navigable areas, and in accordance with the assessment, measures in preventing casualties and improvements of awareness and safety can be suggested and applied. In order to improve security of inland waterways, a review of existing protecting measures of inland waterways, as well as unification of all security rules and legislation is recommended.

Bibliography

Baričević, H. (2001) *Tehnologija kopnenog transporta (Inland transport technology)*, Pomorski fakultet Rijeka, Rijeka.

Bošnjak, I. and Badanjak, D. (2005) *Osnove prometnog inženjerstva (Basic of traffic engineering)*, Fakultet prometnih znanosti, Zagreb.

Čolić, V., Radmilović, Z., and Škiljaica, V. (2005) *Vodni saobraćaj (Inland traffic)*, Univerzitet u Beogradu, Saobraćajni fakultet, Beograd.

Commission of the European Communities (2006) *Communication from the Commission on the Promotion of Inland Waterway Transport "NAIDES*, Brussels. http://ec.europa.eu/transport/iw/doc/2006_01_17_naiades_staff_working_en.pdf.

CRUP (2006) *Priručnik za unutarnju plovidbu u Republici Hrvatskoj (Inland Navigation Manual of Republic Croatia)*, Zagreb.

CRUP and Jolić, N. (2008) *Studija operacionalizacije Nacionalne RIS središnjice RH (studija) (Study of operatization of national centre of RIS of Republic Croatia)*.

Fritell, F.J. (2005) *Port and Maritime Security*, CRS Report for Congress.

Heathcote, P. (2005) *New Measures for Maritime Security Aboard Ships and in Port Facilities.*

IMO (2003) *ISPS Code*

Kasum, J., Marusic, E., and Grzetic, Z. (2006) Security of non-convention ships and nautical tourism ports. *Proceedings of the TIEMS, Seoul, S. Korea*, p. 113.

Kasum, J., Vidan, P., and Baljak, K. (2006) *Act About Safety Protection of Merchant Ships and Ports open to International Traffic and its Implementation, ICTS, Portorož.*

Kasum, J., Baljak, K., and Vidan, P. (2007) *Evaluation of the Existing Piracy Protection Measures*, ISEP, Ljubljana, Slovenia.

Kasum, J., Vidan, P., and Baljak, K. (2007) *A Contribution to Enhancement of Protection Measures Against Piracy*, TIEMS, Trogir.

Kasum, J., Vidan, P., and Baljak, K. (2008) *Threats to Ships and Ports of Inland Navigation*, POWA, Dubrovnik.

Kasum, J., Vidan, P., and Baljak, K. (2010) Threats and new protection measures in inland navigation. *Promet*, **22** (2), 143–146.

Kasum, J., Vidan, P., and Skračić, T. (2010) Maritime Radiation Protection and Seamen's Safety, ISEP, ISEP Proceedings, p. M3, Ljubljana, Slovenija.

Reuter Newswire (1996) USSR and E. Europe. *Reuter Economics News*, January 20.

Slob, W. (1998) Determination of risks on inland waterways. *Journal of Hazardous Materials*, **61**, 363–370.

Vidan, P. (2010) *Model povečanja sigurnosti plovidbe na unutarnjim plovnim putovima(The Model to increase the Safety of Navigation on Inland Waterways)*. Doctoral dissertations. Fakultet prometnih znanosti, Zagreb.

Vidan, P., Kasum, J., and Zujić, M. (2010) Poboljšanje traganja i spašavanja na unutarnjim plovnim putovima (Improvement of search and rescue on inland waterways). *Naše more*, **56** (5–6), 187–192.

Consulted Web Sites

http://ec.europa.eu/transport/iw/index_en.htm (accessed 21 December 2009).

http://epp.eurostat.ec.europa.eu/portal/page/portal/eurostat/home/ (accessed 25 January 2011).

http://www.2.mvr.usace.army.mil (accessed 12 March 2009).

http://www.access.gpo.gov/uscode/title33/title33.html (accessed 12 March 2009).

www.americanwaterways.com (accessed 12 March 2009).

www.CorpsResults.us (accessed 1 February 2009).

www.theodora.com (accessed 25 November 2010).

http://www.trb.org/Conferences/MTS/4A%20GrierPaper.pdf (accessed 12 May 2010).

www.bureauveritas.com (accessed 21 March 2010).

7
Security of Hazmat Transports by Pipeline

Paul W. Parfomak

7.1
Introduction

There are more than a million miles of high volume pipeline transporting natural gas, oil, and other hazardous liquids across the globe.[1] These transmission pipelines are integral to international energy supplies and have vital links to other critical infrastructure, such as power plants, airports, and military bases. While an efficient and fundamentally safe means of transport, these pipelines carry volatile materials with the potential to cause public injury and environmental damage. The world's pipeline networks are also widespread, running alternately through sparsely and densely populated regions; some above ground, some below, some underwater; in politically stable countries and in regions of conflict. They are all vulnerable to theft and sabotage.

This chapter explores the security of pipelines transporting hazardous materials (hazmat) with an emphasis on policy considerations. It describes the global pipeline network and characterizes the nature of security risks to pipelines, including the recent history of international pipeline security incidents. As an example of a coordinated national effort to increase pipeline security, the chapter reviews the US hazmat pipeline sector response to the terror attacks of September 11, 2001. The chapter closes with a discussion of key policy challenges facing governments and the pipeline industry seeking to secure their pipeline systems from hostile actors.

7.1.1
Hazmat Pipeline Infrastructure Around the World

The international network of hazmat transmission pipelines is extensive. High-volume pipelines carrying hazmat commodities are found in 126 countries,

1) Hazardous liquids primarily include crude oil, gasoline, jet fuel, diesel fuel, home heating oil, propane, and butane. Other hazardous liquids transported by pipeline include anhydrous ammonia, carbon dioxide, kerosene, liquefied ethylene, and some petrochemical feedstocks.

Security Aspects of Uni- and Multimodal Hazmat Transportation Systems, First Edition. Edited by Genserik L.L. Reniers, Luca Zamparini.

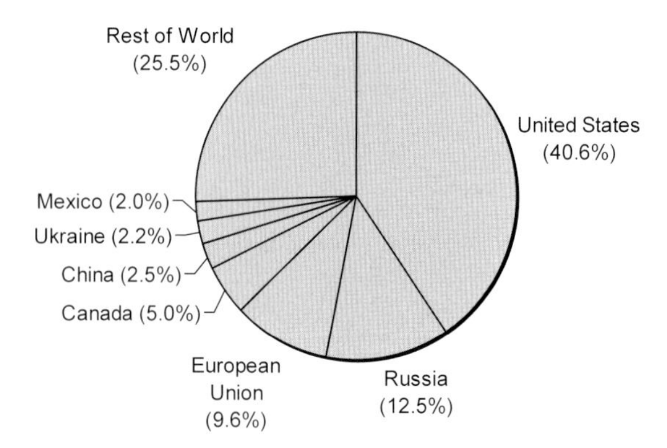

Figure 7.1 Global Distribution of Hazmat Pipeline Mileage. Source: U.S. Central Intelligence Agency, *The World Factbook*, 2010.

dominated by the United States, Russia, the European Union nations, and Canada; the four regions account for over two-thirds of global pipeline mileage (Figure 7.1). (U.S. Central Intelligence Agency, 2010). On the basis of mileage, most hazmat pipelines carry either natural gas (66%) or crude oil (28%), with natural gas almost completely dependent upon pipelines for intra- and international movement (notwithstanding marine shipping of liquefied natural gas). The split of pipeline mileage across commodities varies by country, however, depending upon the nature of their production or consumption of those commodities, and the locations of major sources and markets within their specific transportation networks. China's hazmat pipeline mileage, for instance, is divided relatively evenly between natural gas pipelines on the one hand, and a combination of oil and refined products pipelines on the other (U.S. Central Intelligence Agency, 2010). <Although many major pipelines operate entirely within the borders of a single country, others form the backbone of large and complex networks linking multiple countries or geopolitical regions. The most extensive interconnected networks of hazmat pipelines are in North America and Europe, with the natural-gas system in the latter connecting to nearly every country on the continent as well as to natural gas producers in North Africa and elsewhere (Figure 7.2). Other major pipeline systems are found in the Middle East and, increasingly, Eastern Asia. From a security perspective, this interconnectedness is important because a service interruption anywhere in the network may have significant impacts far downstream. For example, in the Russian–Ukrainian natural gas crisis of 2006, during which Russia briefly cut natural gas supplies to Ukraine, the latter withdrew from transborder pipelines natural gas intended for other European countries. The result was a ripple effect throughout Europe, with Austria, France, Germany, Hungary, Italy, Poland, and Slovakia all reporting pressure drops in their natural gas pipeline systems of around 30% (BBC News, 2006). Conversely, a robust, interconnected pipeline network may minimize the wider impacts of an individual

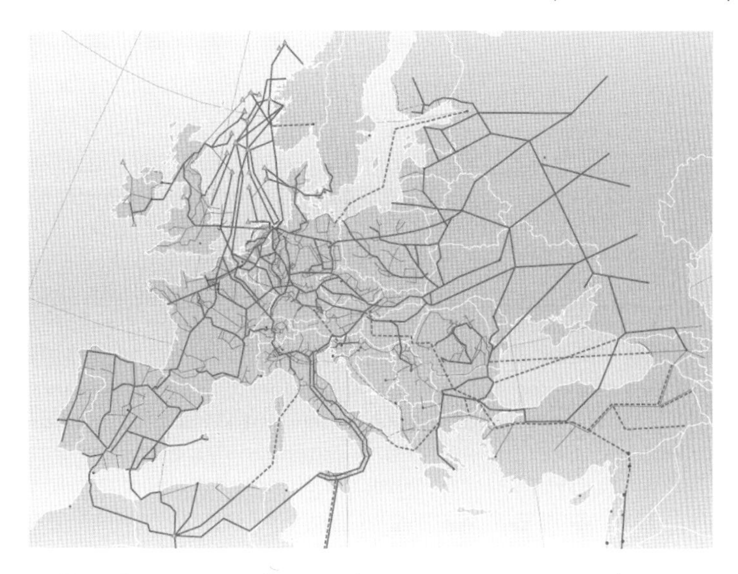

Figure 7.2 Major European Natural Gas Pipelines. Source: Eurogas, *Annual Report 2008-2009*, Brussels, 2010, pp. 34–35.

pipeline incident, because commodities can be rerouted to maintain essential supplies to key delivery points. Pipeline security must therefore be viewed in the context of large, interregional networks, with critical routes and interconnections having a potentially disproportionate influence on the system as a whole. Protecting pipelines from security incidents is simultaneously an issue for governments within their national borders, and for operators, who must ensure the integrity of an entire line across borders.

7.2
Security Risks to Hazmat Pipelines

Hazmat pipelines face two principal categories of security threat: theft and sabotage (typically associated with terrorism). Although they are motivated by different drivers, both theft and sabotage have been key concerns in the pipeline industry throughout the world. Determining whether a particular pipeline security incident falls into one category or another can sometimes be problematic, especially in countries like Nigeria where both economic and political motivations may be at work simultaneously (Houreld, 2007). Nonetheless, theft and sabotage can lead to the same set of consequences, including physical damage to the pipelines, interruption of service, loss of product, public injury, destruction of surrounding property, and environmental damage.

7.2.1
Commodity Theft from Pipelines

There is no comprehensive repository of global information about commodity thefts from hazmat pipelines, but anecdotal statistics for specific operators convey the magnitude of the problem in some pipeline systems. For example, the Nigerian National Petroleum Corporation (NNPC) has reported that, between 2000 and 2010, its pipeline system experienced 15 865 “vandalism” incidents generally involving fuel theft (compared to only 398 releases due to ordinary operational reasons) (Nigerian National Petroleum Corp., 2010). From 2002 to 2009, thieves illegally tapped oil and gas pipelines in China 19 804 times, even including a handful of taps into undersea pipelines in the Shengli oil field (Xie and Economides, 2010). Turkey’s state-owned pipelines reported 411 illegal tapping incidents between 2003 and 2008, driven in part by extremely high refined-product prices in the Turkish market (Alsancak, 2010). In Mexico, drug cartels have become increasingly involved in theft of crude oil from state-owned Petroleos Mexicanos (Pemex) pipelines for sale in the local black market or for export to the United States. Last year, Pemex found 712 illegal connections to its pipeline network siphoning off more than 3.6 million barrels of crude oil. The Mexican drug cartels involved are sophisticated in their approach, employing current and former oil industry workers skilled in pipeline operations, and providing expertise in smuggling stolen oil to US markets using bribes, false transit documents, and legitimate market intermediaries (Campbell, 2011).

Commodity thefts from pipelines are far less common in the United States, Canada, and Western Europe, although they have occurred. In the early 1980s a criminal group tapped into a 16-inch oil pipeline in California, stealing 10 million gallons of crude over a 3-year period before they were caught (Bowers, 2006). Italian authorities report a number of cases of theft or attempted theft from oil pipelines by organized crime over several decades (Pisa, 2008). If global oil and refined product prices remain high, there will be continuing incentive for individuals to steal these commodities from production and transportation facilities in both wealthier and poorer countries.

7.2.2
Global Terrorist Attacks on Pipelines

While theft has been a long-standing security concern for international pipelines, the threat of terrorist attacks has become a more pressing priority in some regions. Energy infrastructure generally, and pipelines in particular, have become favored targets of terrorists groups such as Al Qaeda for several reasons. Pipelines are found in many places, are readily accessible, and are relatively “soft” critical infrastructure targets. Pipeline attacks, although local in nature, also can have greatly amplified economic impacts due to tight supply and demand balances and associated price volatility in major energy markets (Perl, 2008). Attacks on oil and gas pipelines also align with radical Islamist objections to perceived Western “plunder-

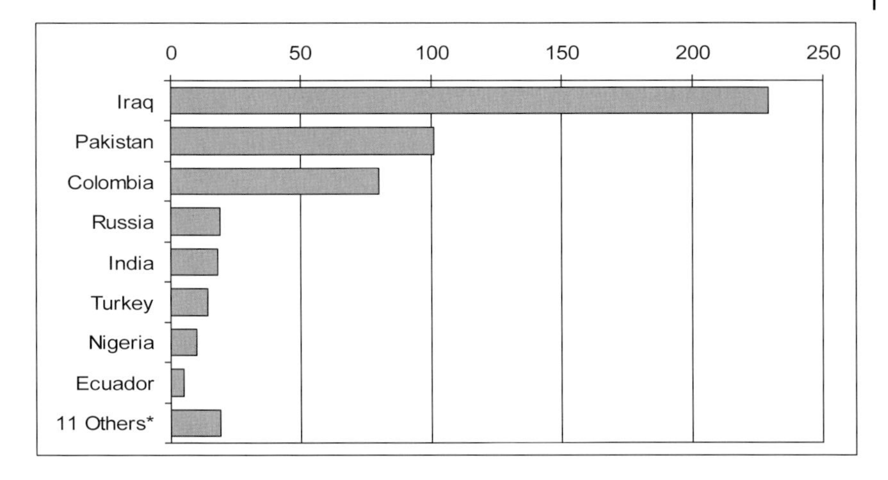

*Afghanistan, Algeria, Chechnya, Georgia, Indonesia, Iran, Myanmar, Nepal, Sudan, Thailand, and Venezuela

Figure 7.3 Global Terrorist Incidents Involving Pipelines 2000–2008. Source: RAND Corp., RAND Database of Worldwide Terrorism Incidents, Online database, 2011, http://www.rand.org/nsrd/projects/terrorism–incidents.html.

ing" of natural resources in the Middle East and North Africa (Giroux, 2009). In populated areas, pipeline attacks can cause spectacular explosions and fires, and can cause significant injuries to people and destruction of property–drawing significant media attention and causing increased concern among the general public. Groups such as Al Qaeda appear to choose the scale and timing of their attacks in order to maximize such media coverage, and hence, public awareness and psychological impact.

The RAND corporation maintains a database of global terrorist incidents, which includes pipeline incidents (RAND Corp., 2011). Although the database appears to exclude several recent terrorist incidents involving North American pipelines, it nonetheless offers helpful perspectives on where and how most terrorist incidents occur around the world. For the period from 2000 through 2008 (the last full year of available data), the RAND database reports 495 terrorist incidents involving pipelines in 19 countries (Figure 7.3). The overwhelming number of terrorist incidents since 2000 have happened in countries experiencing extended military or paramilitary conflict–including Iraq, Pakistan, Colombia, Russia, and Nigeria.[2)] In these countries, many terrorist attacks on pipelines could be extensions of a broader local campaign against a range of civilian and military targets. According to the RAND statistics, the most common threat from terrorist groups is the use of improvised explosive devices to damage hazmat pipelines or related facilities

2) It is interesting to note that Nigeria is reported to have had only 10 "terrorist" attacks on pipeline targets compared to over 15 000 pipeline incidents motivated by theft. Clearly, Nigerian pipeline security is more a matter of preventing crime rather than preventing terrorism.

(e.g., pumping stations). Other studies show that pipelines are also vulnerable to attacks on computer control systems or attacks on electricity grids and communications networks that support pipeline operations (Shreeve, 2006). A series of "coordinated covert and targeted cyberattacks . . . against global oil, energy, and petrochemical companies" in 2009 and 2010 to collect proprietary company operating and financial information have highlighted such computer system vulnerabilities (McAfee Foundstone Professional Services and McAfee Labs, 2011).

A casual examination of the RAND pipeline terrorism statistics might lead one to conclude that protecting pipelines against terrorism is really a significant issue only in Asian and South American conflict zones. However, the rise of radical Islamist terrorist groups following the attacks on the United States of September 11, 2001 and the Madrid train bombings of 2004 also heightened general concerns about terrorist attacks on pipelines in Western countries. As the U.S. Department of Homeland Security stated in its 2008 pipeline sector threat assessment,

> "Transnational terrorist groups, such as Al-Qa'ida, pose the primary threat to the U.S. pipeline network. Single-issue extremist groups, industry insiders, and lone wolves may also pose a threat to the pipeline system . . . U.S.- or Western-owned/operated pipelines in foreign countries are at increased risk of attack by Islamic extremists who oppose a foreign presence or who see a foreign presence as an embezzlement of natural resources." (U.S. Dept. of Homeland Security, Transportation Security Administration, 2011)

A European Parliament report in 2009 likewise concluded,

> "Terrorist attacks at the individual oil or gas pipeline may cause significant local and global damage . . . Although energy supply chains in Europe have so far not been targeted, the threat of oil supply disruptions is real and the risks are growing." (European Parliament, 2009)

Assessments by government authorities in Canada and other Western countries have reached similar conclusions about the recent rise in terrorist threats to energy pipeline infrastructure (Gendron, 2010, p.13).

Over the last 15 years, a number of terrorist incidents among pipelines in North America and Europe, involving both Islamist and non-Islamist actors, appear to validate a heightened security posture in the hazmat pipelines sector. In 1996, London police foiled a plot by the Irish Republican Army to bomb gas pipelines and other utilities across the city (President's Commission on Critical Infrastructure Protection, 1997). A Mexican rebel group detonated bombs along Mexican oil and natural gas pipelines in July and September, 2007 (Johnson, 2007). In June, 2007, the U.S. Department of Justice arrested members of a terrorist group planning to attack jet-fuel pipelines and storage tanks at the John F. Kennedy International Airport in New York (U.S. Dept. of Justice, 2007). Natural gas pipelines in British Columbia, Canada, were bombed six times between October 2008 and July

2009 by unknown perpetrators in acts classified by authorities as environmentally motivated "domestic terrorism" (Gelinas, 2010).

One Western pipeline of particular concern, and with a history of terrorist and vandal activity, is the Trans-Alaska Pipeline System (TAPS) in the United States, which transports crude oil from Alaska's North Slope oil fields to a marine terminal in Valdez for offloading onto waterborne oil tankers. TAPS runs some 800 miles and delivers nearly 17% of United States domestic oil production (Alyeska Pipeline Service Co., 2009). In 1999, Vancouver police arrested a man planning to blow up TAPS for personal profit in oil futures (Cloud, 1999). In 2001, a vandal's attack on TAPS with a high-powered rifle forced a two-day shutdown and caused extensive economic and ecological damage (Rosen, 2002). In January 2006, federal authorities acknowledged the discovery of a detailed posting on a website purportedly linked to Al Qaeda that reportedly encouraged attacks on US pipelines, especially TAPS, using weapons or hidden explosives (Loy, 2006). In November 2007 a US citizen was convicted of trying to conspire with Al Qaeda to attack TAPS and a major natural gas pipeline in the eastern United States (U.S. Attorney's Office, 2007). The most recent US government threat assessment concludes that, notwithstanding the expressed interest of transnational terrorists in attacking pipeline infrastructure, the terrorist threat to U.S pipelines is low (U.S. Transportation Security Administration, 2011). Nonetheless, pipeline operators in the United States and elsewhere remain alert.

7.2.3
Costs and Impacts of Pipeline Security Incidents

From an economic perspective, pipeline security incidents can be costly due to repair and response costs, commodity losses, curtailment of transportation service to shippers, and commodity market impacts. Total costs to Nigeria's NNPC associated with pipeline thefts over the last ten years have amounted to over US$1.14 billion and have rendered key pipelines such as the Trans-Forcados Pipeline inoperable for months at a time (Nigerian National Petroleum Corp., 2010). Pemex losses in 2008 from illegal pipeline taps were US$750 million (Gould and Rodriguez, 2010). Rebel bombings of Colombia's Caño Limón oil pipeline in 2001–2002 reportedly caused $500 million in lost oil production, equivalent to 2 per cent of the Colombian national budget (Robles, 2002a). From 2002 to 2006, thefts from oil and gas pipelines in China caused losses of $72 million to the China National Petroleum Corp. (Xin, Li, and Wan, 2010). Thefts from its oil pipelines in the Russian Caucasus region cost Transneft approximately $20 million in 2008 (Bierman, 2009). The aggregate economic impact of security incidents in other countries may be lower, but even one incident, depending upon its location and severity, can lead to damages and loss of product costing millions of dollars.

In addition to their economic impacts, attacks of hazmat pipelines can be destructive to life and property. Explosions and fires from the pipeline incidents in Nigeria between 2000 and 2010 killed over 1200 people, including 260 deaths in a single fire in 2006 (Bello, 2010). Pipeline thefts in Mexico have also led to serious accidents, such as a 2010 explosion at the Nuevo Teapa pipeline that killed 28 people and damaged 115 homes (Gould and Rodriguez, 2010). In 2006, approximately 63 Iraqis were killed, another 140 injured, and 4 buildings destroyed when insurgent bombs ruptured a natural gas pipeline in Baghdad (U.S. Central Command, 2006). A 1998 National Liberation Army bombing of the Ocensa oil pipeline in Colombia killed 84 people and injured at least 100 others, destroying much of the village of Machuca (Salgado, 2007). These are among the most serious pipeline incidents in recent years in terms of human casualties, but even a more limited human impact may serve the objectives of terrorists, especially in more developed countries. As one study concluded, "to make it into the news, terrorists operating in Western countries can commit some minor terror incident with few fatalities" (Frey and Rohner, 2006). Accordingly, terrorists attacking pipelines in the United States or Europe may achieve their media objectives even with relatively limited casualties and environmental impacts. Extensive US press coverage of a 2010 natural gas pipeline *accident* in California which killed 8 people, injured 60 others, and destroyed 37 homes offers an indication of the heavy media coverage given to fatal pipeline incidents in the United States.

7.2.4
Responding to Pipeline Security Threats

In the face of security threats, hazmat pipeline operators and government agencies have implemented a range of measures for preventing pipeline security incidents – including intelligence gathering, facilities hardening, increasing system redundancy, adding surveillance, and upgrading computer systems to prevent cyber attacks. The nature and emphasis of these measures has varied depending upon the particular security situation faced by specific pipelines. Most operators put in place at least a minimum set of physical security measures, such as barriers and access controls at critical facilities, security cameras, security patrols, and other basic measures (many also serving essential safety functions). Among developing countries with a history of extensive thefts or pipeline attacks, it has been common to deploy the military or large private security forces to physically patrol pipelines and deter saboteurs. In Iraq, for example, some 14 000 security guards were posted at one time along pipelines and associated installations (Luft, 2005). Likewise, Occidental Petroleum historically has paid the Colombian army to protect its frequently attacked, 480-mile long Caño Limón pipeline in regions dominated by Colombian rebels (Robles, 2002b). Some operators have also put in place advanced pipeline monitoring technologies, such as impact monitors and fiber-optic sensors, to detect illegal taps and other physical disturbances, although these types of technologies have significant limitations and can be circumvented (Hopkins, 2008). Many pipeline operators have also made substantial investments in computer software to protect their supervisory control and data acquisition (SCADA) systems against viruses, hackers, and other cyber threats (Dickman, 2010). In addition to deterring security incidents, pipeline operators also try to reduce the impacts of pipeline attacks through improved incident response and the development of alternative, redundant transportation routes.

There are few publicly available sources of information about the direct costs to hazmat pipeline operators of pipeline security measures. In conflict regions security costs may be relatively large. For example, between 2002 and 2005, the US government provided the Colombian government with $99 million in aid to help secure the Caño Limón pipeline against rebel attacks (Government Accountability Office, 2005). The funds were used for helicopters, equipment, logistical and infrastructure support, and training–in addition to security measures already in place and paid for by the pipeline's owner and the Colombian government. For some pipelines in less-developed countries, operators or other stakeholders may also be required to pay (sometimes, bribe) local authorities or potential saboteurs to forestall pipeline thefts and attacks (Mroue, 2003). In developed countries where pipeline security incidents are infrequent, however, security costs may be relatively small. For example, in a 2006 tariff filing with federal regulators, one US oil pipeline operator sought to recover US$0.008 per barrel of transported oil to cover costs incurred for additional pipeline security measures put in place after 9/11 such as additional security cameras, lighting, fencing, and gates (Rocky Mountain Pipeline System LLC, *Local Pipeline Tariff Applying on Petroleum Products as Defined in Item No. 10 Transported by Pipeline from and to Points Named Herein*, Filing with the U.S. Federal Energy Regulatory Commission, FERC No. 157, Issued May 31, 2006). This surcharge amounted to less than 1% of the average rate charged for oil transportation on the pipeline, an almost negligible surcharge on the total delivered cost of a barrel of crude.

7.3 US Pipeline Security after September 11, 2001

The United States has, by far, the largest network of hazmat transmission pipelines in the world. Although almost all of these pipelines are owned and operated by private companies, the federal government regulates the safety of hazmat pipelines and has a significant role in ensuring establishing and overseeing pipeline security programs. It is, therefore, instructive to review the measures taken by both pipeline operators and US government agencies to bolster pipeline security in the aftermath of the September 11, 2001 terror attacks. In particular, policy decisions and security programs of the federal government offer a well-developed example of ways that government agencies and pipeline operators can work together to protect pipelines from security threats.

7.3.1 Pipeline Operator Security Programs

US pipeline operators have always sought to secure their pipeline assets, but their security programs historically tended to focus on personnel safety and preventing vandalism. In a few specific cases, security has been more comprehensive with more of an eye towards organized attacks. For example, security at the trans-Alaska pipeline during the Persian Gulf War of 1990–1991 included measures such as armed guards, controlled access, intrusion detection, dedicated communications at key facilities, and both aerial and ground surveillance of the pipeline corridor (U.S. General Accounting Office, 1991). However, the events of 9/11 focused attention on the vulnerability of critical infrastructure, including pipelines, to different terrorist threats. In particular, the terrorist attacks raised the possibility of systematic attacks on pipelines by sophisticated radical Islamic terror groups in a manner that had not been widely contemplated before. After the 9/11 attacks, both hazmat pipeline operators and government agencies in the United States immediately increased security and began identifying additional ways to deal with increased terrorist threats.

Gas pipeline operators in the United States, through one of their main trade associations, formed a security task force to coordinate and oversee the industry's security efforts. Among its first priorities was to ensure, among other things, that every member company designated a senior manager to be responsible for security. Working with the federal Department of Transportation (DOT), the Department of Energy (DOE), and nonmember pipeline operators, the gas pipeline association assessed existing gas pipeline industry security programs and began developing common risk-based practices for incident deterrence, preparation, detection, and recovery. These assessments addressed issues such as spare parts exchange, critical parts inventory systems, and security communications with emergency agencies, among other matters. Gas pipeline operators also worked with federal agencies, including the DOT and the newly established federal Office of Homeland Security, to develop a common government threat notification system (Haener and CMS Energy Corp., 2002).

The natural gas companies reported significant commitments to bolster security at their critical facilities. Companies strengthened emergency, contingency and business continuity plans; increased liaison with law enforcement; increased monitoring of visitors and vehicles on pipeline property; monitored pipeline flows and pressure on a continuous basis; increased employee awareness of security concerns; and deployed additional security personnel (American Gas Association [AGA], 2002). The industry also began developing encryption protocol standards to protect gas systems from cyber attacks (Ryan, 2002). Operators also sought redundancy in the delivery system to provide greater flexibility to redirect or shut down product flows.

The oil pipeline industry responded to the September 11 attacks in a manner similar to that of the natural gas pipeline industry. Oil and refined product pipeline operators reviewed procedures, tightened security, rerouted transportation patterns, closely monitored visitors, and made capital improvements to harden key facilities (Shea, 2002). Operators also increased surveillance of pipelines, conducted more thorough employee background checks, and further restricted Internet mapping systems (Association of Oil Pipelines [AOPL], 2003). The oil pipeline and oil industry trade associations, working together, provided guidance to member companies on how to develop a recommended pipeline security protocol analogous to an existing protocol on managing pipeline safety. Along with the gas pipeline industry, the oil pipeline industry reconciled its levels of security threat and associated measures with the national threat advisory system of the Office of Homeland Security. By February, 2003, 95 per cent of oil pipeline operators had developed new security plans and had instituted the appropriate security procedures. The remaining 5 per cent were primarily small operators in other businesses, with oil pipelines between plant facilities (Association of Oil Pipelines [AOPL], 2003).

In conjunction with the Office of Homeland Security, pipeline operators joined with other gas and oil companies to establish an Energy Information Sharing and Analysis Center (EISAC) in November 2001. The center, which remains active, is a cooperative, industry-directed database and software applications center for information related to security, including real-time threat alerts, cyber alerts and solutions. It allows authorized individuals to submit reports about information and physical security threats, vulnerabilities, incidents, and mitigation. The EISAC also provides access to information from other industry members, US government, law enforcement agencies, technology providers, and other security associations (Energy Information Sharing and Analysis Center, 2003).

7.3.2
Security Initiatives of the U.S. Department of Transportation

Presidential Decision Directive 63 issued by President Bill Clinton in 1998 assigned lead responsibility for pipeline infrastructure protection to the DOT. (Presidential Decision Directive 63, 1998) At the time, these responsibilities fell to the DOT's safety office, since that office was already addressing some elements of pipeline security in its role as safety regulator. Immediately after September 11, 2001, the DOT issued several emergency bulletins to oil and gas pipeline companies communicating the need for a heightened state of alert in the industry. Soon thereafter, because of national security concerns, the DOT removed from its Internet web site detailed maps of the country's pipeline infrastructure that had previously been available to viewing by the general public. The DOT also conducted a vulnerability assessment used to identify which pipeline facilities were "most critical" because of their importance to meeting national energy demands or proximity to highly populated or environmentally sensitive areas. The agency worked with industry groups and state pipeline safety organizations ". . . to assess the industry's readiness to prepare for, withstand and respond to a terrorist attack . . ." (U.S. Department of Transportation, 2001). The DOT warned that critical pipeline facilities, such as control centers, pump and compressor stations, and storage facilities, might be targets – and that many of these facilities needed to be better protected (U.S. Department of Transportation, 2003).

Through 2002, the DOT was the federal agency most active in encouraging industry activities intended to better secure US pipelines. In general, the agency's approach was to encourage operators to voluntarily improve their security practices rather than to develop new security regulations. In adopting this approach, the agency sought to speed adoption of security measures by industry and avoid the publication of sensitive security information (e.g., critical facility lists) that would normally be required in public rulemaking (U.S. Government Accounting Office, 2002). The agency also worked with several industry security task groups to define different levels of pipeline asset "criticality," to identify actions to strengthen protection based on this criticality, and to develop plans for improved response preparedness. The DOT surveyed many pipeline companies to assess security measures taken since 9/11. Together with the DOE and state pipeline agencies, the DOT promoted the development of consensus standards for security measures tiered to correspond with the five levels of terrorism threat warnings issued by the Office of Homeland Security (Engleman, 2002a). The DOT also developed protocols for inspections of critical facilities to ensure that operators implemented appropriate security practices. To convey emergency information and warnings, the agency established a variety of communication links to key staff at the most critical pipeline facilities throughout the country. The DOT also began identifying near-term technology to enhance deterrence, detection, response and recovery, and began seeking to advance public and private sector planning for response and recovery (Engleman, 2002b).

On September 5, 2002, the DOT circulated formal guidance defining the agency's security program recommendations and implementation expectations. This guidance recommended that operators identify critical facilities, develop security plans consistent with prior trade association security guidance, implement security plans and review those plans annually (O'Steen, 2003). The guidance defined asset "criticality" in terms of threats, risks to people, and economic impacts from the loss of energy supply. It also suggested specific security measures to be taken at the different homeland security threat levels, with over 50 cumulative measures at the highest threat level. While the guidance was voluntary, the DOT expected compliance and informed operators of its intent to begin reviewing security programs within 12 months, potentially as part of more comprehensive safety inspections. The federal pipeline security authority was subsequently transferred outside of the DOT, however, as discussed below, the agency did not follow through on a national program of pipeline security program reviews.

7.3.3 Transportation Security Administration and Pipeline Security

In November 2001, President George W. Bush signed the Aviation and Transportation Security Act establishing the Transportation Security Administration (TSA) within the DOT. The act placed the DOT's pipeline security authority within TSA including a range of duties and powers related to general transportation security, such as intelligence management, threat assessment, mitigation, security measure oversight, and enforcement, among others. On November 25, 2002, President Bush signed the Homeland Security Act of 2002 creating the Department of Homeland Security (DHS). Among other provisions, the act transferred to DHS the Transportation Security Administration from the DOT.

On December 17, 2003, President Bush issued Homeland Security Presidential Directive 7 clarifying executive agency responsibilities for identifying, prioritizing, and protecting critical infrastructure.[3] The directive maintains DHS as the lead agency for pipeline security and instructs the DOT to "collaborate in regulating the transportation of hazardous materials by all modes (including pipelines)." The directive also requires that DHS and other federal agencies collaborate with "appropriate private sector entities" in sharing information and protecting critical infrastructure. TSA joined both the Energy Government Coordinating Council and the Transportation Government Coordinating Council. (The missions of the councils are to work with their industry counterparts to coordinate critical infrastructure protection programs in the energy and transportation sectors, respectively, and to facilitate the sharing of security information.) The directive also required DHS to develop a national plan for critical infrastructure and key resources protection, which the agency issued in 2006 as the *National Infrastructure Protection Plan*. That plan, in turn, required each critical infrastructure sector to develop a sector-specific plan that describes strategies to protect its critical infrastructure, outlines a coordinated approach to strengthen its security efforts, and determines appropriate funding for these activities. Executive Order 13416 further required the transportation sector-specific plan to prepare annexes for each mode of surface transportation (Executive Order 13416, 2006). In accordance with the above requirements the TSA issued its *Transportation Systems Sector Specific Plan* and *Pipeline Modal Annex* in 2007.

According to the agency's *Pipeline Modal Annex*, TSA has been engaged in a number of specific pipeline security initiatives since 2003, as summarized in Table 7.1.

In 2003, TSA initiated its Corporate Security Review (CSR) program, wherein the agency visits the largest pipeline and natural gas distribution operators to review their security plans and inspect their facilities. During the reviews, TSA evaluates whether each company is following the intent of the voluntary security guidance, and seeks to collect the list of assets each company had identified meeting the criteria established for critical facilities. In 2004, the DOT reported that the plans reviewed to date (approximately 25) had been "judged responsive" to the agency's initial pipeline security guidance (Department of Transportation [DOT], 2004). As of August 2010, TSA had completed CSRs covering the largest 100 pipeline systems (84% of total US energy pipeline throughput) and was in the process of conducting second CSRs of these systems (Government Accountability Office, 2010b). Recent results indicate that the majority of US pipeline systems "continue to do a good job in regards to pipeline security" although there are areas in which pipeline security can be improved (Transportation Security Administration, 2010a). Past reviews have identified inadequacies in some company security programs such as not updating security plans, lack of management support, poor employee involvement, inadequate threat intelligence, and employee apathy or error (Gillenwater, 2007). In 2008, the TSA initiated its Critical Facility Inspection Program, under which the agency conducts indepth inspections of all the critical facilities of the 100 largest pipeline systems in the United States. By 2012, TSA

3) HSPD-7 supersedes PDD-63.

Table 7.1 TSA pipeline security initiatives.

Initiative	Description	Participants[a]
Pipeline policy and planning	Coordination, development, implementation, and monitoring of pipeline security plans	TSA, DHS, DOT, DOE
Sector coordinating councils and joint sector committee	Government partners coordinate interagency and crossjurisdictional implementation of critical infrastructure security	TSA, DOE, other agencies, industry
Corporate security reviews (CSR)	Onsite reviews of pipeline operator security	TSA, industry
Pipeline system risk tool	Statistical tool used for relative risk ranking and prioritizing CSR findings	TSA, industry
Pipeline crossborder vulnerability assessment	US and Canadian security assessment and planning for critical crossborder pipeline	TSA, Canada
Regional gas pipeline studies	Regional supply studies for key natural-gas markets	TSA, DOE, INGAA, GTI, NETL, industry
Cyber attack awareness	Training/presentations on Supervisory Control and Data Acquisition (SCADA) system vulnerabilities	TSA, GTI
Landscape depiction and analysis tool	Incorporates depiction of the pipeline domain with risk analysis components	TSA
International pipeline security forums	International forums for US and Canadian governments and pipeline industry officials convened annually	TSA, Canada, other agencies, industry
"G8" multinational security assessment and planning	Multinational sharing of pipeline threat assessment methods, advisory levels, effective practices, and vulnerability information; also develops a G8-based contingency planning guidance document	TSA, DHS, state dept., G8 nations
Pipeline security drills	Facilitation of pipeline security drills and exercises	TSA, industry
Security awareness training	Informational compact discs about pipeline security issues and improvised explosive devices	TSA
Stakeholder conference calls	Periodic information-sharing conference calls between key pipeline security stakeholders	TSA, other agencies, industry
Pipeline blast mitigation studies	Explosives tests on various pipe configurations to determine resiliency characteristics	TSA, DOD, other agencies
Virtual library pipeline site	Development of TSA information-sharing Web portal	TSA

Sources: Transportation Security Administration, Pipeline Modal Annex, June 2007, pp. 10–11; Jack Fox, Transportation Security Administration, Testimony before the House Committee on Homeland Security, Subcommittee on Management, Investigations, and Oversight, April 19, 2010.

a) Key: DHS = Dept. Of Homeland Security, DOE = Dept. of Energy, G8 = Group of Eight (US, UK, Canada, France, Germany, Italy, Japan, and Russia), GTI = Gas Technology Institute, INGAA = Interstate Natural Gas Association of America, NETL = National Energy Technology Laboratory, TSA = Transportation Security Administration.

expects to complete inspections for all 373 critical facilities identified by pipeline operators (GAO, 2010).

In addition to the initiatives in Table 7.1, TSA has worked to establish qualifications for personnel applying for positions with unrestricted access to critical pipeline assets and has developed its own inventory of critical pipeline infrastructure (Transportation Security Administration, 2003). The agency has also addressed legal issues regarding recovery from terrorist attacks, such as Federal Bureau of Investigation control of crime scenes and eminent domain in pipeline restoration. In October 2005, TSA issued an overview of recommended security practices for pipeline operators "for informational purposes only . . . not intended to replace security measures already implemented by individual companies" (Transportation Security Administration, Intermodal Security Program Office 2005). The agency released revised guidance on security best practices at the end of 2006, and again in 2011. The guidelines include a section on cybersecurity developed with the assistance of the Applied Physics Laboratory of Johns Hopkins University as well as other government and industry stakeholders (Transportation Security Administration, 2010b).

The pipeline mission of the TSA currently includes developing security standards; implementing measures to mitigate security risk; building and maintaining stakeholder relations, coordination, education and outreach; and monitoring compliance with security standards, requirements, and regulations. The TSA's pipeline security program has traditionally received from the DHS's general operational budget an allocation for routine operations such as regulation development, travel, and outreach. The budget funds 13 full-time equivalent staff within the office in 2011 (Transportation Security Administration, 2010c).

In 2007 the TSA Administrator testified before Congress that the agency intended to conduct a pipeline infrastructure study to identify the "highest risk" pipeline assets, building upon such a list developed through the CSR program. He also stated that the agency would use its ongoing security review process to determine the future implementation of baseline risk standards against which to set measurable pipeline risk-reduction targets (Hawley, 2007). Provisions in the Implementing Recommendations of the 9/11 Commission Act of 2007 require TSA, in consultation with the DOT, to develop a plan for the federal government to provide increased security support to the "most critical" pipelines at high or severe security alert levels and when there is specific security threat information relating to such pipeline infrastructure. The act also requires a recovery protocol plan in the event of an incident affecting the interstate and intrastate pipeline system.

In addition to the above pipeline security initiatives, TSA has performed a limited number of vulnerability assessments and has supported investigations for specific companies and assets where intelligence information has suggested potential terrorist activity. The agency, along with the DOT, was involved in the investigation of an August, 2006 security breach at a liquefied natural gas (LNG) peak-shaving plant in Massachusetts (U.S. Department of Transportation, 2006). Although not a terrorist incident, the security breach involved the penetration of intruders through several security barriers and alert systems, permitting them to access the main LNG storage tank at the facility. The TSA also became aware of the JFK airport terrorist plot in its early stages and supported the Federal Bureau of Investigation's associated investigation. The agency engaged the private sector in helping to assess potential targets and determine potential consequences. The agency worked with the pipeline company to keep it informed about the plot, discuss its security practices, and review its emergency response plans (Transportation Security Administration, 2007).

In August 2010, the Government Accountability Office (GAO) released a report examining TSA's efforts to ensure pipeline security. The report focused on TSA's use of risk assessment and risk information in securing pipelines, actions the agency has taken to improve pipeline security under guidance in the 9/11 Commission Act of 2007, and the agency's efforts to measure such security improvement efforts (Government Accountability Office, 2010a). Among other findings, the report concluded that, although TSA had begun to implement a risk-management approach to prioritize its pipeline security efforts, work remained to ensure that the highest risk pipeline systems would get the necessary scrutiny. The report also concluded that TSA was missing opportunities under its inspection programs to better ensure that pipeline operators understand how they can enhance the security of their pipeline systems. GAO found that linking TSA's pipeline security performance measures and milestones to the goals and objectives in its national security strategy for pipeline systems could aid in achieving results within specific time frames and could facilitate more effective oversight and accountability (Government Accountability Office, 2010b). TSA concurred with all of the report's recommendations for addressing the issues and is in the process of implementing them (Levine, 2010).

7.3.4
Federal Energy Regulatory Commission

One area related to pipeline safety and security not under either the DOT's or TSA's primary jurisdiction is the siting approval of new natural-gas pipelines, which is the responsibility of the Federal Energy Regulatory Commission (FERC). Companies building interstate natural gas pipelines must first obtain from FERC certificates of public convenience and necessity. (The commission does not oversee oil pipeline construction.) FERC must also approve the abandonment of gas facility use and services. These approvals may include safety and security provisions with respect to pipeline routing, safety standards and other factors, although, as a practical matter, FERC has traditionally left these considerations to the other agencies.

In 2001, FERC notified jurisdictional companies that it would "approve applications proposing the recovery of prudently incurred costs necessary to further safeguard the nation's energy systems and infrastructure" in response to the terror attacks of 9/11. The commission also committed to "expedite the processing on a priority basis of any application that would specifically recover such costs from wholesale customers." Companies could propose a surcharge over currently existing rates or some other cost recovery method (Federal Energy Regulatory Commission [FERC], 2001). In FY2005, the commission processed security cost-recovery requests from 14 oil pipelines and 3 natural gas pipelines (Federal Energy Regulatory Commission [FERC], 2006). FERC's FY2006 annual report stated that "the Commission continues to give the highest priority to deciding any requests made for the recovery of extraordinary expenditures to safeguard the reliability and security of the Nation's energy transportation systems and energy supply infrastructure" (Federal Energy Regulatory Commission [FERC], 2007).

In 2003, FERC promulgated a new rule to protect "critical energy infrastructure information," defined as information that "must relate to critical infrastructure, be potentially useful to terrorists, and be exempt from disclosure under the Freedom of Information Act." According to the rule, critical infrastructure is "existing and proposed systems and assets, whether physical or virtual, the incapacity or destruction of which would negatively affect security, economic security, public health or safety, or any combination of those matters." Critical energy infrastructure information excludes "information that identifies the location of infrastructure." The rule also establishes procedures for the public to request and obtain such critical information, and applies both to proposed and existing infrastructure (Federal Energy Regulatory Commission [FERC], 2003). In 2003, FERC also handed down new rules facilitating the restoration of pipelines after a terrorist attack–allowing owners of a damaged pipeline to use blanket certificate authority to immediately start rebuilding, regardless of project cost, even outside existing rights-of-way. Pipeline owners would still need to notify landowners and comply with environmental laws. Prior rules limited blanket authority to $17.5 million projects and 45-day advance notice (Schmollinger, 2003).

7.4 Policy Issues in Hazmat Pipeline Security

Over the past decade, facing increased threats of theft and terrorist attack, hazmat pipeline companies and government agencies around the world have taken substantial actions to improve pipeline security and oversight. As one US pipeline company executive remarked, "Before 9/11, we never contemplated somebody flying a jet into some critical facility we have. Now we not only think about it, but we're also putting very different contingency plans in place." (Field & People's Energy Corp., 2002). For the most part, these actions appear to be reducing the frequency of pipeline security incidents overall. Terrorist attacks against pipelines worldwide, especially in Iraq and Pakistan, have fallen off sharply since the mid-2000s. Commodity thefts from pipelines are also down from their recent peaks in Nigeria, Mexico, and Turkey. Nonetheless, international hazmat pipeline systems continue to face key policy issues associated with threat information, identifying critical facilities, and international cooperation, among other issues.

7.4.1
Security Threat Information

Among the greatest challenges to hazmat pipeline system operators is the availability of current and actionable security threat information in a global environment where security threats are constantly changing. Especially in the case of terrorists threats, pipeline operators seek site-specific threat information to make better security decisions and focus protection where it is truly needed. But such threat information is often not readily available, either because the intelligence systems have not been established to collect and analyze it, or because government authorities who have energy-sector threat information do not appropriately share it with pipeline companies. (Perl, 2008, p. 4) Lacking threat intelligence at the site level, pipeline operators must "make a threat assumption that international terrorism is possible at every facility that has adequate attractiveness to that threat" (American Petroleum Institute, 2005). By uniformly responding to ambiguous warnings in this way, hazmat pipeline operators may expend significant resources to increase security at facilities that are not really under increased threat. The timely development and sharing of security information among pipeline operators and government intelligence agencies is, therefore, essential to deterring pipeline attacks and maximizing the protective value of pipeline security investments.

7.4.2
Identifying Critical Pipeline Facilities

Given the physical scale of most hazmat transmission pipeline networks, identifying critical facilities within those networks is important for prioritizing security measures. But the definition of "critical" is often ambiguous, often varies from one jurisdiction to another, and may change depending upon the context in which it is used. Compilation of critical pipeline infrastructure lists also poses confidentiality concerns, since such lists could be useful to competitors for commercial purposes or to hostile actors seeking attractive pipeline targets. Consequently, identifying critical pipeline assets has been problematic for many pipeline operators. The United States, for example, has a history of missteps and ambiguity identifying critical pipeline infrastructure, initially avoiding numerical thresholds and relying instead on discretionary qualitative metrics like "significance" of impact (American Gas Association and Interstate Natural Gas Association of America, 2002). A 2010 report by the Canadian Department of Defense concluded that "(n)o national critical infrastructure assets or services have yet been identified or agreed with the various owners/operators and federal, provincial and territorial authorities" even though there is a legal imperative to protect such infrastructure (Gendron, 2010, p. 34). The European Union issued a directive as "a first step in a step-by-step approach to identify and designate (European critical infrastructures) and assess the need to improve their protection" only in December 2008 (Council of the European Union, 2008). Efforts in developing countries (e.g., India) appear to be further behind those in wealthier nations, although they may require the same dedicated attention in the context of national infrastructure security initiatives (Srivastava, 2009).

7.4.3 International Cooperation in Pipeline Security

As global demand for energy grows, the global hazmat pipeline network will grow with it, expanding existing multinational pipeline connections and establishing new ones. This trend will drive an ever-greater imperative for international cooperation to secure critical global pipeline links (Yergin, 2006). As the head of the antiterrorism unit of the Organization for Security and Co-operation in Europe has stated,

> "many actors of the energy sector feel that as the energy infrastructure system is transnational, a need exists for international efforts towards development of a uniform cross-border regulatory framework and comprehensive set of international standards for energy infrastructure security" (Perl, 2008).

The head of the Anti-Terrorist Center of the Commonwealth of Independent States has similarly concluded, "it is essential to also focus on cooperation among States that supply, transit and receive energy resources, including strengthening counterterrorist collaboration among intelligence agencies" (Novikov, 2010). But achieving increased cooperation in pipeline security among neighboring countries could pose a daunting set of institutional, legal, cultural, and economic challenges. For example, should Austria share the costs of government pipeline security initiatives in Turkey for a new natural gas pipeline linking the two countries? Can a group of countries develop a harmonized approach to pipeline vulnerability and threat assessment? There are successful examples of bilateral government cooperation in key areas of hazmat pipeline security, such as between the United States and Canada, but it remains to be seen whether effective multinational programs for pipeline security can be instituted more broadly around the world.

7.5 Conclusions

This brief review of international hazmat pipeline security suggests that "pipeline security" varies as an issue depending upon the geographic region and socioeconomic environment in which the term is applied. In zones of armed conflict, the principal pipeline security concern may be systematic bombings associated with a broader insurgent or paramilitary campaign against the government or occupation forces. In poorer oil producing countries, pipeline security may be primarily associated with theft of pipeline products – sometimes by local opportunists, sometimes by organized criminals. In Western countries, pipeline security may be most strongly associated with isolated acts of Islamist or environmental terrorism. These differences are important in that they may lead to different strategies for protecting pipelines and responding to security incidents. For example, it would make little sense for European nations to enlist their armies to protect their pipeline networks, as Mexico and Colombia have done. Conversely it is unlikely that a sophisticated, Islamic-focused intelligence effort as practiced by the United States, would be of much value in Nigeria for preventing oil thefts by unorganized villagers seeking fuel for their trucks. Ultimately, the nature of hazmat pipeline security must be considered country by country, and pipeline by pipeline, to most effectively match security measures to very specific assessments of regional pipeline threats. As international hazmat pipeline security evolves, having the best information and effective international coordination will be the keys to maximizing the benefits of limited pipeline security resources.

Bibliography

Alsancak, H. (2010) The role of Turkey in global energy: bolstering energy infrastructure security. *Journal of Energy Security*.

Alyeska Pipeline Service Co. (2009) *Internet page, Anchorage, AK, February 8, 2009.* http://www.alyeska-pipe.com/?about.html (accessed November 6, 2009).

American Gas Association (AGA) (2002) *Natural Gas Distribution Industry Critical Infrastructure Security*, AGA, Natural Gas Infrastructure Security–Frequently Asked Questions, April 30, 2003.

American Gas Association and Interstate Natural Gas Association of America (2002) *Security Guidelines: Natural Gas Industry Transmission and Distribution*, Washington, DC. September 6, 2002.

American Petroleum Institute (2005) *Security Guidelines for the Petroleum Industry*, Washington, DC, April 2005, p. 5.

Association of Oil Pipelines (AOPL) (2003) *Protecting Pipelines from Terrorist Attack*, In the Pipe, Washington, DC, February 10, 2003.

BBC News (2006) *Q&A: Ukraine Gas Row*, January 4.

Bello, O. (2010) *Pipeline Vandalism Still Threatens Human Lives, Economy*, Business Day, Lagos, Nigeria, September 29, 2010.

Bierman, S. (2009) *Transneft Fights 27 000-Ton Oil Theft as Violence Surges*, Bloomberg, September 1,2009.

Bowers, A. (2006) *Slate's Explainer: Stealing Natural Gas from a Pipeline*, National Public Radio, January 3, 2006.

Campbell, R. (2011) *Mexico Oil Thieves Get Sophisticated Amid Crackdown, Reuters*, January 20, 2011.

Cloud, D.S. (1999) *A Former Green Beret's Plot to Make Millions Through Terrorism*, Ottawa Citizen, December 24, 1999, p. E15.

Council of the European Union (2008) *Council Directive 2008/114/EC.*

Department of Transportation (DOT) (2004) *Action Taken and Actions Needed to Improve Pipeline Safety*, CC-2004-061, June 16, 2004, p. 21.

Dickman, F. (2010) *Enhancing SCADA and Pipeline Security*, Pipeline and Gas Technology, April 1, 2010.

Energy Information Sharing and Analysis Center (2003) *About the Energy ISAC.* Internet home page. Washington, DC. May 2003.

Engleman, E. (2002a) Administrator, U.S. Department of Transportation, *Research and Special Programs Administration, Statement before the U.S. House of Representatives Committee on Energy and Commerce*, Subcommittee on Energy and Air Quality, March 19, 2002.

Engleman, E. (2002b) Administrator, U.S. Department of Transportation, *Research and Special Programs Administration,*

Statement before the U.S. House of Representatives Committee on Transportation and Infrastructure, Subcommittee on Highways and Transit, February 13, 2002.

European Parliament (2009) *Directorate-General for Internal Policies, Gas and Oil Pipelines in Europe*, PE 416.239 (IP/A/ITRE/NT/2009-13), p. 19.

Executive Order 13416, Strengthening Surface Transportation Security, December 5, 2006.

Federal Energy Regulatory Commission (FERC) (2001) *News release, R-01-38*, Washington, DC, September 14, 2001.

Federal Energy Regulatory Commission (FERC) (2003) *News release, R-03-08*, Washington, DC. February 20, 2003.

Federal Energy Regulatory Commission (FERC) (2006) *Federal Energy Regulatory Commission Annual Report FY2005*, p. 19. These are the most recent specific figures reported.

Federal Energy Regulatory Commission (FERC) (2007) *Federal Energy Regulatory Commission Annual Report FY2006*, p. 23.

Field, D.; People's Energy Corp. (2002) *Chicago, IL, as quoted in: Karen Ryan, "Powerful Protection," American Gas*, Washington, DC, May, 2002.

Frey, B.S. and Rohner, D. (2006) *Blood and Ink! The Common-Interest-Game Between Terrorists and the Media*. Center for Research in Economics, Management, and the Arts, Basel, Switzerland, Working Paper No. 2006-8, p. 18.

Gelinas, B. (2010) *New Letter Threatens Resumption of "Action" against B.C. Pipelines*, Calgary Herald, April 15, 2010.

Gendron, A. (2010) *Critical Energy Infrastructure Protection in Canada, Defence R&D Canada, Centre for Operational Research & Analysis, DRDC CORA CR 2010-274.*

Gillenwater, M. (2007) *TSA, Pipeline Security Overview*, Presented to the Alabama Public Service Commission Gas Pipeline Safety Seminar, Montgomery, AL, December 11, 2007.

Giroux, J. (2009) *Targeting Energy Infrastructure: Examining the Terrorist Threat in North Africa and its Broader Implications*, Real Instituto Elcano, Madrid, February 13, 2009.

Gould, J.E. and Rodriguez, C.M. (2010) *Pemex Pipeline Blast Blamed on Criminals Kills 28*, Bloomberg, December 20, 2010.

Government Accountability Office (2005) *Security Assistance: Efforts to Secure Colombia's Caño Limón-Coveñas Oil Pipeline Have Reduced Attacks, but Challenges Remain*, GAO-05-971, September 2005, p. 2.

Government Accountability Office (2010a) *GAO Watchdog, Transportation Security's Efforts To Ensure Pipeline Security, Assignment No. 440768*, Internet database, February 4, 2010.

Government Accountability Office (GAO) (2010b) *Pipeline Security: TSA Has Taken Actions to Help Strengthen Security, but Could Improve Priority-Setting and Assessment Processes*, GAO-10-867, August, 2010, Executive summary.

Haener, W.J.; CMS Energy Corp. (2002) *Testimony on behalf of the Interstate Natural Gas Association of America (INGAA) before the U.S. House of Representatives Committee on Transportation and Infrastructure*, Subcommittee on Highways and Transit, February 13, 2002, p. 4.

Hawley, K. (2007) Asst. Secretary, Dept. of Homeland Security, *Testimony before the Senate Committee on Commerce, Science, and Transportation hearing on Federal Efforts for Rail and Surface Transportation Security*.

Hopkins, P. (2008) *Learning from Pipeline Failures*, WTIA/APIA Welded Pipeline Symposium, Perth, Australia, March 2008.

Houreld, K. (2007) *Militants Say 3 Nigeria Pipelines Bombed*, Associated Press, May 8, 2007.

Johnson, R. (2007) Six Pipelines Blown Up in Mexico, *Los Angeles Times*, September 11, 2007. p. A–3.

Levine, J.E. (2010) Director, Departmental GAO/OIG Liaison Office, U.S. Dept. of Homeland Security, *Letter to GAO*, July 23, 2010; Transportation Security Administration, Pipeline Security Division, personal communication, November 5, 2010.

Loy, W. (2006) *Web Post Urges Jihadists to Attack Alaska Pipeline*, Anchorage Daily News, January 19, 2006.

Luft, G. (2005) *Pipeline Sabotage is Terrorist's Weapon of Choice*, Pipeline and Gas Journal, February 2005.

McAfee Foundstone Professional Services and McAfee Labs (2011) *Global Energy Cyberattacks: "Night Dragon"*, White paper, February 10, 2011.

Mroue, B. (2003) *Oil Ministry and U.S. Troops Take Measures to Protect Iraq's Main Pipeline from Thieves and Saboteurs*, Associated Press, July 3, 2003.

Nigerian National Petroleum Corp. (2010) *How Pipeline Vandals Cripple Fuel Supply–NNPC. Incurs Over N174 Billion in Products Losses, Pipeline Repairs*, March 3, 2010, http://www.nnpcgroup.com/PublicRelations/NNPCinthenews/tabid/92/articleType/ArticleView/articleId/68/How-Pipeline-Vandals-Cripple-Fuel-Supply–NNPCIncurs-over-N174-billion-in-products-losses-pipeline-repairs.aspx (accessed February 15, 2011).

Novikov, A. (2010) *Protection of Critical Energy Infrastructure against Terrorist Attacks in the Commonwealth of Independent States (CIS), CTN Newsletter Special Bulletin*, Organization for Security and Co-operation in Europe, Action Against Terrorism Unit, January 2010.

O'Steen, J.K. (2003) Deputy Associate Administrator For Pipeline Safety, U.S. Department of Transportation, *Research and Special Programs Administration, Implementation of RSPA Security Guidance*, Presentation to the National Association of Regulatory Utility Commissioners, February 25, 2003.

Perl, R.F. (2008) *Protecting Critical Energy Infrastructures Against Terrorist Attacks: Threats, Challenges and Opportunities for International Co-operation*, Reinforced NATO Economic Committee Meeting, Brussels, September 22, 2008.

Pisa, N. (2008) *Oil Siphon Thief Sparks Fatal Blast*, Sky News, October 2 2008.

Presidential Decision Directive 63 (1998) *Protecting the Nation's Critical Infrastructures.*

President's Commission on Critical Infrastructure Protection (1997) *Critical Foundations: Protecting America's Infrastructures, Washington, DC, October 1997.*

RAND Corp. (2011) *RAND Database of Worldwide Terrorism Incidents, Database Definitions*, Web page, http://www.rand.org/nsrd/projects/terrorism-incidents/about/definitions.html (accessed March 5, 2011). RAND defines terrorist incidents as violent, politically motivated incidents against civilian targets intended to create alarm and coerce certain actions.

Robles, F. (2002a) *U.S. Trains Colombians to Protect Oil Pipeline*, Miami Herald, December 13, 2002.

Robles, F. (2002b) *Pipeline Patrol: U.S. May Pump Aid Into Colombia's Effort To Guard Oil*, Miami Herald, March 16, 2002.

Rocky Mountain Pipeline System LLC, *Local Pipeline Tariff Applying on Petroleum Products as Defined in Item No. 10 Transported by Pipeline from and to Points Named Herein*, Filing with the U.S. Federal Energy Regulatory Commission, FERC No. 157, Issued May 31, 2006.

Rosen, Y. (2002) *Alaska Critics Take Potshots at Line Security*, Houston Chronicle, February 17, 2002.

Ryan, K. (2002) *Powerful Protection*, American Gas. Washington, DC, May, 2002.

Salgado, C. (2007) *ELN Cupola: 40 Years for Machuca*, El Colombiano, March 8, 2007.

Schmollinger, C. (2003) *FERC OKs Emergency Reconstruction*, Natural Gas Week, May 13, 2003.

Shea, W.H. (2002) President and CEO, Buckeye Pipe Line Co. *Testimony on behalf of the Association of Oil Pipe Lines and the American Petroleum Institute before the U.S. House of Representatives Committee on Transportation and Infrastructure*, Subcommittee on Highways and Transit, February 13, 2002.

Shreeve, J.L. (2006) *Science & Technology: The Enemy Within*, The Independent. London, UK, May 31, 2006, p. 8.

Srivastava, L. (2009) *Complexities of Energy Security, Financial Chronicle*, Mumbai, January 27, 2009.

Transportation Security Administration (2003) *TSA Multi-Modal Criticality Evaluation Tool*, TSA Threat Assessment and Risk Management Program, slide presentation, April 15, 2003.

Transportation Security Administration, Intermodal Security Program Office

(2005) *Pipeline Security Best Practices*, p. 1.

Transportation Security Administration (2007) Jack Fox, personal communication.

Transportation Security Administration (2010a) Jack Fox, personal communication, November 5, 2010.

Transportation Security Administration (2010b) Jack Fox, personal communication, February 2, 2010.

Transportation Security Administration (2010c) *Pipeline Security Division*, Jack Fox, personal communication, November 5, 2010.

U.S. Attorney's Office (2007) *Middle District of Pennsylvania, "Man Convicted of Attempting to Provide Material Support to Al-Qaeda Sentenced to 30 Years' Imprisonment," Press release, November 6, 2007*; A. Lubrano and J. Shiffman, "Pa. Man Accused of Terrorist Plot," Philadelphia Inquirer, February 12, 2006, p. A1.

U.S. Central Command (2006) *2 Explosions Rupture Gas Line, Destroys 4 Buildings; Blasts Kill 63 Iraqis, Wound 140*, News release, Baghdad, Iraq, August 13, 2006.

U.S. Central Intelligence Agency (2010) *The World Factbook*, Field Listing: Pipelines, Internet table, https://www.cia.gov/library/publications/the-world-factbook/fields/2117.html?countryName=Nigeria&countryCode=ni®ionCode=af&#ni (accessed March 5, 2011).

U.S. Dept. of Homeland Security, Transportation Security Administration (2011) *Pipeline Threat Assessment*, October 23, 2011, p.3.

U.S. Dept. of Justice (2007) *Four Individuals Charged in Plot to Bomb John F. Kennedy International Airport*, Press release, June 2, 2007.

U.S. Department of Transportation (2001) *Research and Special Programs Administration*, RSPA Pipeline Security Preparedness, December, 2001.

U.S. Department of Transportation (2003) *Research and Special Programs Administration*, RSPA Budget Estimates: Fiscal Year 2003, p. 106.

U.S. Department of Transportation (2006) *Pipeline and Hazardous Materials Safety Administration, "Pipeline Safety: Lessons Learned From a Security Breach at a Liquefied Natural Gas Facility," Docket No. PHMSA-04-19856, Federal Register*, Vol. 71, No. 249, December 28, 2006, p. 78269; TSA, Intermodal Security Program Office, personal communication, August 30, 2006.

U.S. General Accounting Office (1991) *Trans-Alaska Pipeline: Ensuring the Pipeline's Security*, GAO/RCED-92-58BR, Washington, DC, November, 1991, p. 12.

U.S. Government Accounting Office (2002) *Pipeline Security and Safety: Improved Workforce Planning and Communication Needed*, GAO-02-785, August, 2002, p. 22.

U.S. Transportation Security Administration (2011) *Pipeline Threat Assessment*.

Xie, X. and Economides, M.J. (2010) *The Growing Problem of Oil Theft in China*, *Energy Tribune*, June 9, 2010.

Xin, D., Li, J., and Wan, Z. (2010) *China Faces New Risk: Attacks on Pipelines*, China Daily, January 6, 2010.

Yergin, D. (2006) *Ensuring Energy Security*, *Foreign Affairs*, Vol. 85 No. 2, March 1, 2006, p. 78.

Part Three
Security of Hazmat Transports: Multimodal Perspectives

8
Mulitmodal Transport: Historical Evolution and Logistics Framework

Wout Dullaert, Bert Vernimmen, and Luca Zamparini

8.1
Introduction

Since the early days of intermodal, multimodal and combined transport the academic literature has offered a variety of definitions. Faust (1985) defines multimodal transport as "the transport of goods by at least two different modes of transport on the basis of a single multimodal transport contract". Hayuth (1987) explains intermodality as "the movement of cargo from shipper to consignee by at least two different modes of transport under a single rate, through-billing, and through-liability. The objective of intermodal transportation is to transfer goods in a continuous flow through the entire transport chain, from origin to final destination, in the most cost- and time-effective way". Nowadays, the European Union, the European Conference of Ministers of Transport (ECMT) and the United Nations Economic Commission for Europe (UN/ECE) have accepted the following definition of combined transport in Europe: "Intermodal transport where the major part of the journey, in Europe, is by rail, inland waterways or sea, and any initial and/or final legs carried out by road are as short as possible." (UIRR, 2011).

Both in academic and in professional publications, intermodal, multimodal and combined transport are often used as synonyms. They all share the same difference from unimodal transport: several transport modes are used for moving cargo from an origin to a destination, usually involving stopovers at terminals for changing transport modes. Most intercontinental transport is multimodal as seagoing vessels or aircraft are usually unable to deliver directly to the final customer. From seaports or airports the cargo can be shipped to the final customer by road, rail, barge/short sea vessel or pipeline.

Multimodal or combined transport offers a wide variety of potential advantages to direct road transport such as moving larger volumes at lower freight rates during a significant part of the transport chain, the possibility to circumvent bans on the circulation of heavy commercial vehicles in Europe (e.g., on Sundays in France and Germany), etc. When comparing different transport alternatives, one should however, not only consider the transportation cost of each alternative, but

Security Aspects of Uni- and Multimodal Hazmat Transportation Systems, First Edition. Edited by Genserik L.L. Reniers, Luca Zamparini.

also the inventory costs associated with each option. The total logistics framework pioneered by Baumol and Vinod (1970) has proven to be a fruitful approach to supporting operational decision making and boast a large body of literature that considers a variety of key model extensions that are often not covered in standard logistics management textbooks and, as a result, are little known in practice.

Yet, when applying the basic framework to real-life cases, the true cost of a transport alternative can be seriously under- or overestimated. In this chapter we will illustrate that this is *a forteriori* true for multimodal transport by presenting the basic framework and two key extensions why they are of particular interest when considering multimodal transport: the measurement of the service level and the shape of the demand during lead-time distribution.

The remainder of this chapter is structured as follows. Section 8.2 discusses the evolution of multimodal transport in the European Union, in the United States and in Asia. Section 8.3 presents the case study that we are going to use to discuss the standard framework for comparing transport alternatives. This methodology is explained in Section 8.4 and the case is re-examined in Section 8.5 to illustrate the impact of service-level measurement and demand during lead-time distributions. Section 8.6 concludes the chapter.

8.2 Evolution of Multimodal Transport in the European Union, in the United States and in Asia

The present section will provide some considerations on multimodal transport in three of the most relevant regions of the world (the European Union, the United States and ASEAN countries). In the cases of the European Union and of the United States, it will provide some statistical data on the evolution and on the composition of multimodal transport, while in the case of ASEAN countries, the main issues related to the development of this sector will be presented and discussed.

8.2.1 Multimodal Transport in the European Union

When European multimodal transport is taken into account, a reliable source of statistics and time series is represented by the International Union of combined road–rail transport companies (UIRR). Although over time additional companies have joined the UIRR, the amount of cargo handled by its 18 member companies in 12 European countries in Table 8.1 suggests a steadily growing adoption of multimodal transport (see European Union, 2010 for more details).

An analysis of the time series suggests that the largest increase in multimodal transport has been related to international shipments (that have passed from 12 billion ton-km in 1990 to 35.7 billions of ton-km in 2009). On the other hand, the increase in the amount of national combined transport has been much more

Table 8.1 Evolution of multimodal transport in the European Union.

UIRR companies					
Year	**Billions of tonne-kilometres**		**Traffic**		
	(share of overall freight transport in brackets)		**% of consignments**		
		of which: national	**Semitrailers**	**Rolling road**	**Swap bodies**
1990	19.0	7.0	20%	18%	62%
1995	25.0	7.3	14%	19%	67%
1996	27.2	7.6	12%	20%	68%
1997	29.9	8.3	10%	19%	71%
1998	30.2	8.3	9%	20%	71%
1999	28.6	7.8	9%	22%	69%
2000	32.5	8.2	9%	23%	68%
2001	31.9	7.2	9%	24%	67%
2002	33.1	8.0	8%	23%	69%
2003	32.9	7.7	7%	23%	70%
2004	34.5	8.3	7%	16%	67%
2005	37.0	8.1	7%	13%	80%
2006	45.4	9.8	9%	16%	76%
2007	46.1	9.8	9%	14%	77%
2008	46.0	10.3	8%	14%	78%
2009	46.0	10.3	8%	14%	78%

Source: UIRR.

limited (passing from 7 billion ton-km to 10.3 billion ton-km). This may be partially due to geographical reasons but may also depend on the regulations of some countries (i.e. Ecopoints) that foster the adoption of transport modes alternative to road haulage to pass through them.

Furthermore, it is striking that the accompanied rolling road traffic is losing market share compared to unaccompanied rail/road traffic. When semitrailers or swap bodies are used, a different driver will pick up the cargo at the destination terminal to make the final delivery to the customer. This may be due to the largest role played by multimodal transport operators that oversee the entire series of transportation activities required by a multimodal shipment, hiring difference local transport firms for the initial and final legs of a shipment.

A research conducted by Blauwens, Vernimmen, and Witlox (2003) with annual data spanning from 1992 to 2001 pointed to the opposite conclusion on the evolution of unaccompanied traffic, suggesting a recent shift since 2003. The share of semitrailers within unaccompanied transport has more or less consistently dropped in favor of swap bodies since 1990.

Table 8.2 Evolution of multimodal transport in the United States (1993–2007); share of multimodal with respect to overall freight transport in brackets.

	Value (Billions$)	Tons (Millions)	Ton-km (Billions)
1993	663(11.34)	226(2.33)	191(7.88)
1997	946(13.62)	217(1.95)	205(7.70)
2002	1079(12.84)	217(1.85)	226(7.20)
2007	1867(15.97)	574(4.57)	417(12.46)

Source: US Department of Transportation 2009.

8.2.2
Multimodal Transport in the United States

The evolution and the expected future developments of multimodal transport in the United States can be analyzed by using the data provided by the US Department of Transportation. In a publication that gave a synthesis of the 2007 Commodity Flow Survey (US Department of Transportation, 2009), multimodal transport was considered both in terms of tons and ton-km and value shipped (Table 8.2). It must be stressed that according to all three measures considered, there has been a remarkable increase in the amount of multimodal transport both in absolute values and as a percentage of the overall freight shipments. The comparison between the value and the tons shipped shows that firms recurring to multimodal transport produce and/or trade goods that have a density value that is much higher than the average density value given that, for example, in 2007, the quantity moved by multimodal transport was 4.57% of the overall quantity, while its value was close to 16% of the overall value.

On the other hand, the comparative analysis in 2007 (but the same holds true for the other sampled years) of the shipped tons (which, as mentioned above, represented 4.57% of the overall quantity) and the typical freight transport measure represented by the ton-km (which accounted for 12.46% of the total) shows, as expected, that the average distance in multimodal transport is much higher than in the case of unimodal transport.

The same US Department of Transportation study compared the percentage change in ton-miles from 1993 to 2007 (Figure 8.1) for all accounted transport alternatives and for the overall freight-transport sector. It clearly emerged that multimodal transport has experienced the largest increase in the considered time period. It has grown by 118%, which represents more than twice the percentage growth of road haulage (54%) and almost three times the growth in rail freight (43%), the transport alternative that accounts for the largest share of freight shipments in the United States.

A further publication of the US Department of Transportation (2010) has attempted to estimate the very long run future scenarios of freight transport in the United States (Table 8.3) and has given the tentative dimension of this sector in 2040.

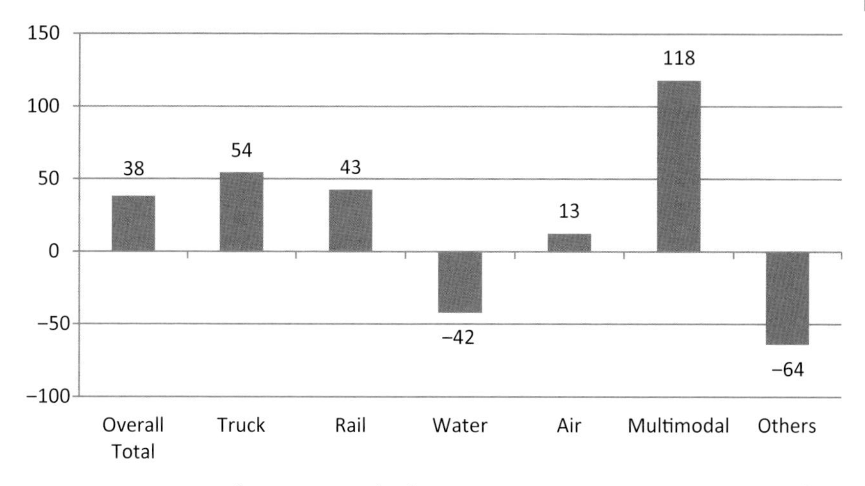

Figure 8.1 Percentage change in ton-miles from 1993 to 2007. Source: US Department of Transportation 2009.

Table 8.3 Expected evolution of volumes and values of domestic freight shipments in US.

	Volume (Billions of tons)		Value (Billions of 2007$)	
	2009	**2040**	**2009**	**2040**
Total	14.397	22.772	12.078	29.444
Truck	10.713	17.963	9.087	20.114
Rail	1.575	2.109	323	477
Water	351	482	99	128
Air, air & truck	3	5	147	740
Multiple modes & mail[1]	458	724	1.618	6.728
Pipeline	1.069	1.158	532	585
Other & unknown	229	331	273	672

Source: US Department of Transportation 2010.

Table 8.3 indicates that the overall shipped quantities should increase by 58% passing from 14.397 billion tons to 22.772 billion tons. In the same period, the increase in multimodal transport in the United States should follow a very similar pattern and increase by 58%. However, when the value of the goods is taken into account, the overall increase in value should equal 143%, while the increase in value of the goods shipped through multimodal transport should be equal to 315%. This implies that the relative value density of goods shipped through multimodal transport should increase.

8.2.3 Multimodal Transport in the ASEAN Countries

An important agreement for multimodal transport in South East Asia was issued, under the aegis of ASEAN, in October 2000. In a study by Goh *et al.* (2008), it emerges that the air and maritime transport infrastructures in the region are of an adequate and homogeneous standard. On the other hand, projects are underway in order to have an efficient railway network in the area. However, problems appear to characterize the interconnectivity among modes. Until recently, among the 45 largest ports in the region, only 13 have connectivity with railways and only 9 with inland waterways.

Another important issue for multimodal transport development is represented by crossborder infrastructure investments that should not only foster transport activities but also economic cooperation. These investments should allow reduction of the variance in the main logistics indicators and in the global infrastructure competitiveness that are shown in Table 8.4 (Bhattacharyay, 2010). It is apparent that some of these countries (Malaysia and Thailand, not to mention Singapore that is not present in the table but that ranks 4th in the world for infrastructure competitiveness) offer multimodal operators an environment that can forge the development of this sector, while most other countries are lagging behind. In this respect, the Philippines appear as the country that implies the longest lead time to move goods from the shipper to the port of loading or from the port of unloading to the consignee while Myanmar displays the longest spell for average customs clearance time. In order to provide an efficient transport network for multimodal transport, the ASEAN countries have envisaged an overall investment of 450 billion US$ in the coming decade, 317 billion US$ should be devoted to the provi-

Table 8.4 Logistics indicators in selected ASEAN countries (2008).

Country	Customs clearance (average in days)	Lead time export, median case (days)	Lead time import, median case (days)	Infrastructure competitiveness score (rank out of 134 countries)
Cambodia	1.00	2.71	3.29	3.1(82)
Indonesia	1.58	2.54	3.88	3.0(86)
Malaysia	1.68	3.44	3.31	5.3(23)
Myanmar	4.48	2.65	3.16	–
Philippines	1.82	6.35	5.31	2.9(92)
Thailand	1.92	3.39	2.29	4.7(29)
Vietnam	1.45	2.77	3.95	2.7(93)

Lead time is related to the amount of time necessary to move goods from the shipper to port of loading (export) or from the port of unloading to consignee (import).
Source: Bhattacharyay, 2010.

sion of increased capacity, while 113 billion US$ will be needed for the maintenance of the existing infrastructure.

The present section has discussed the evolution of multimodal transport in some of the most relevant economic regions indicating the importance of value density, interconnectivity and lead times. The next sections will provide a tentative model that includes a wide variety of logistics cost factors and that can provide an explanation of the choice among multimodal options for (inter)continental transport.

8.3 Problem Statement

Vernimmen (2004) presents the following case of a company located in Dortmund, Germany that regularly imports products from the US East Coast. It will be used as a starting point to illustrate the potential and the limitations of the inventory-theoretic approach to modal choice for supporting multimodal transport decision making.

The products are shipped by container from the port of New York to the port of Antwerp by sea. On an annual basis, the freight flow represents 182 container loads, each having an average value of 50 000 EUR. On average, half a container load is consumed per day, with a variance of 0.1 container load per day^2.

For the transport from Antwerp to Dortmund, the shipper has three options: unimodal road transport, combined rail–road transport and combined barge-road transport. In the latter cases, containers are first shipped to Duisburg in Germany by either rail or barge (286 or 285 km, respectively) from which they are loaded on trucks to reach the final destination at 50 km from the port of Duisburg. The cost of inventory during transport and in the warehouse, the so-called holding rate, is, respectively, 20% and 25% per annum. The key data on product demand and logistics parameters are summarized in Table 8.5.

8.4 The Standard Framework

The transport rates of the multimodal transport options are significantly lower than that of road transport. Rail/road is 26% cheaper than direct road transport and the barge/road option is even up to 35% cheaper. Both multimodal transport options are, however, characterized by a longer average transport time that also has a higher variance.

The competitive freight rates for multimodal transport are partially explained by the fact that the port of Duisburg is located at a significant distance from the port of Antwerp, which enables barge and rail to offset the additional container handling costs (compared to direct road transport) by lower transportation costs. In fact, oncarriage from the port of Duisburg to Dortmund only covers 50 km but

Table 8.5 Product and transport-mode characteristics.

Product value C in EUR	50 000					
Annual demand R in containers	182					
Cycle service level	0.99					
Avg demand per day	0.5					
Var demand	0.1					
Mode	freight rate	distance in km	avg lead time	var lead time	frequency/ week	shipment quantity
Sea transport New York to Antwerp			12	1	7	
Road transport to Dortmund	420	240	0.2	0.02		1
Rail transport to Duisburg	140	286	0.5	0.1	5	1
Road transport Duisburg–Dortmund	170	50	0.1	0		
Barge transport to Duisburg	100	285	1	0.1	3	1
Road transport Duisburg–Dortmund	170	50	0.1	0		

Table 8.6 Composition of the overall lead time form New York to Dortmund by rail/road transport.

Departures per week	Sea transport		Waiting time before departure		Rail transport to Duisburg		Road transport to Dortmund		Overall lead time rail transport	
	avg	var	avg	var	avg	var	avg	var	**avg**	**var**
7	12.00	1.00	0.50	0.08	0.50	0.10	0.10	0.00	**13.10**	**1.18**
6	12.00	1.00	0.58	0.11	0.50	0.10	0.10	0.00	**13.18**	**1.21**
5	12.00	1.00	0.70	0.16	0.50	0.10	0.10	0.00	**13.30**	**1.26**
4	12.00	1.00	0.88	0.26	0.50	0.10	0.10	0.00	**13.48**	**1.36**
3	12.00	1.00	1.17	0.45	0.50	0.10	0.10	0.00	**13.77**	**1.55**
2	12.00	1.00	1.75	1.02	0.50	0.10	0.10	0.00	**14.35**	**2.12**
1	12.00	1.00	3.50	4.08	0.50	0.10	0.10	0.00	**16.10**	**5.18**

costs 170 EUR per container. Barge transport from Antwerp to Duisburg over a distance of 285 km, including container handling costs only costs 100 EUR per container.

Hinterland barge/road transport takes more time than the rail/road option for two reasons. First, the operational speed of barge transport is much lower than for rail transport. A train can cover the 286 km distance from Antwerp to Duisburg in around 9 h whereas a barge needs more than a day for a similar distance (partially due to the fact that the barge is sailing against the current). Secondly, the lower frequency of barge transport (three sailings per week between Antwerp and Duisburg, compared to five departures per week for rail) leads to longer lead times and higher variances of waiting time (and hence of the overall lead time) as explained below.

The average transport times and transport-time variance for each transport leg can be easily observed in practice and are independent of the frequency. Sea transport between New York and Antwerp takes 12 days on average with a variance of 1 day^2. Barge transport from Antwerp to Duisburg requires on average 1 day, with a variance of 0.1 day^2. The rail connection to Duisburg is on average 0.5 days faster, with a variance of 0.1 day^2. Oncarriage from the port of Duisburg to the final destination takes 0.10 days for both multimodal alternatives. Given the short distance, the variance is assumed to be zero. In the case of direct road transport from Antwerp to Dortmund, the containers can be delivered fairly quickly and reliably (0.2 days average lead time with a variance of 0.02 days^2).

The waiting time between each of the legs depends on the frequency and can be approximated by assuming that arrival of the transport mode of each leg is uniformly distributed, as in Vernimmen (2004). For the ease of exposition, consider the rail connection with a frequency of 7 departures per week in Table 8.6. Given that the time between two consecutive departures at days a and b is 1 day, a shipment arriving in the port of Antwerp to be sent by rail has to wait on average for 0.5 days $\frac{1}{2}(a+b)$ with a variance of 0.08 day^2 $\left(\frac{(b-a)^2}{12}\right)$.

Along the same lines, the waiting time for a barge departure (with three sailings per week from Antwerp to Duisburg) is on average 1.17 days with an average of 0.45 days^2. Transportation time by barge is assumed to be 1 day with a variance of 0.1 days^2. By assuming that travel times and waiting times for the different legs of the supply chain are independent, the overall average lead time and variance of lead time can be easily calculated for different frequencies, as detailed below (Table 8.7).

A straightforward application of the standard inventory-theoretic approach could consist of determining the transportation costs and the costs of cycle stock, inventory in transit and safety stock for a given risk of a stockout per replenishment cycle per container.

Transportation costs are usually very easy to calculate. Estimating the inventory costs associated with each transport mode is slightly more complicated.

Table 8.7 Composition of the overall lead time form New York to Dortmund by barge/road transport.

Departures per week	Sea transport		Waiting time before departure		Barge transport to Duisburg		Road transport to Dortmund		Overall lead time barge transport	
	avg	var	avg	var	avg	var	avg	var	**avg**	**var**
7	12.00	1.00	0.50	0.08	1.00	0.10	0.10	0.00	**13.60**	**1.18**
6	12.00	1.00	0.58	0.11	1.00	0.10	0.10	0.00	**13.68**	**1.21**
5	12.00	1.00	0.70	0.16	1.00	0.10	0.10	0.00	**13.80**	**1.26**
4	12.00	1.00	0.88	0.26	1.00	0.10	0.10	0.00	**13.98**	**1.36**
3	12.00	1.00	1.17	0.45	1.00	0.10	0.10	0.00	**14.27**	**1.55**
2	12.00	1.00	1.75	1.02	1.00	0.10	0.10	0.00	**14.85**	**2.12**
1	12.00	1.00	3.50	4.08	1.00	0.10	0.10	0.00	**16.60**	**5.18**

Cycle Stock Costs Since the shipment size between Antwerp and Dortmund equals one container, on average half a container will be in stock. With a container load value of EUR 50 000 and warehousing costs of 25%, the cycle stock costs amount to 6250 EUR per year. Since the annual demand is 182 container loads, this results in cycle stock costs of EUR 34.34 per container. It does not differ between the transportation alternatives.

Costs of Inventory In-Transit If direct road transport is used, a container spends on average 12.2 days (12 days sea transport plus 0.2 days for hinterland transport) in the supply chain. With an in-transit holding cost of 20% per year and a value of EUR 50 000 per container, this results in an in-transit inventory cost of 334.25 EUR per container. For barge and rail transport these costs will be higher, given their higher average lead time, especially for lower departure frequencies.

Costs of Safety Stock Safety stock is not related to safety issues such as accidents or injuries in logistics. It is the term used by logisticians to refer to the additional inventory that is maintained to accommodate variability in demand and in the lead time. In the literature, several approaches are discussed to determine safety stocks that broadly fall into one of the following three categories: a “time supply approach”, a “shortage costing approach” or a “service-level approach” (Silver, Pyke, and Peterson, 1998). The time supply approach specifies the safety stock level as a period of demand that it should be able to cover, for example, 1 month. Although it is still a common measure in practice, this approach offers little insight into how often one is able to meet customer demand, that is, the actual service that one is able to offer its customers. As such, it can be considered to be an inferior approach. Under the second approach, the objective is to minimize the total of shortage cost and inventory carrying cost, which will result in a certain amount of safety stock for the inventory item under consideration. The

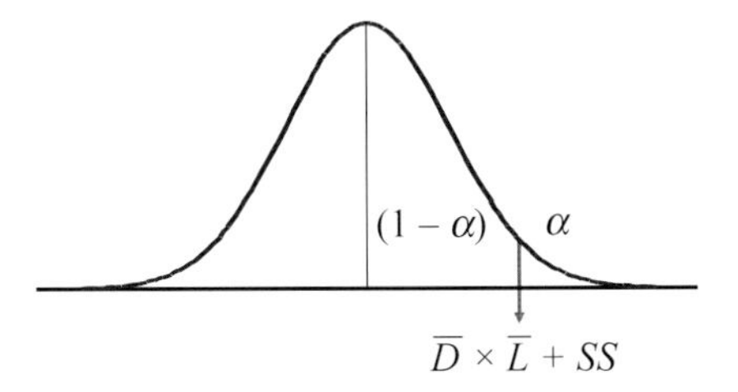

Figure 8.2 Symmetrical demand during lead time.

Achilles' heel of this approach is that practitioners usually find it hard to determine directly how high shortage costs are, as this involves monetarizing qualitative aspects of service such as, e.g., loss of goodwill from customers. This problem is not encountered in the service-level approach, where the objective is to minimize the inventory carrying cost for satisfying a customer demand during a certain percentage of the replenishment cycles (cycle service level) or a certain prespecified percentage of demand (fill rate).

As the calculation of the safety stock based on the cycle service level can be easily derived from the standard normal distribution, it allows for a quick estimate of the required safety stock, as detailed below.

When any order is placed, it takes a while before the order arrives. This time interval is called the lead time and it is usually characterized by its average $\bar{L}$ and variance $\sigma_{\bar{L}}^2$. For each day during the lead time, the company has to face a demand with an average of $\bar{D}$ and variance $\sigma_{\bar{D}}^2$. Most companies will therefore place orders in time to make sure that they can at least cover the expected (or average) demand they will face during the lead time. Figure 8.2 illustrates that when the resulting demand during lead-time distribution is symmetrical, placing an order at a reorder level equal to $\bar{D} \times \bar{L}$ would result in a cycle service level of 50% as there is only 50% probability that the demand during the lead time would be lower than the reorder level. To keep a higher service level of $(1-\alpha)$ per cent, the order should be placed when the inventory level is equal to $(1-\alpha)$ percentile of the demand during lead-time distribution. This reorder level ROP is equal to the average demand during the lead time $\bar{D} \times \bar{L}$ plus the safety stock SS.

By standardizing the values of the demand during lead-time distribution, the amount of safety stock SS can be written as

$$\mathrm{SS} = k\sigma_{\mathrm{DDLT}} \tag{8.1}$$

in which k is the safety factor that specifies how many times the standard deviation of demand during lead time has to be kept as safety stock. When demand and lead-time distributions are independent and identically distributed, the standard deviation of demand during lead time can be calculated as (Silver, Pyke, and Peterson, 1998):

$$\sigma_{\mathrm{DDLT}} = \sqrt{\bar{L}\sigma_{\bar{D}}^2 + \bar{D}^2\sigma_{\bar{L}}^2} \tag{8.2}$$

Table 8.8 Impact of frequency on total logistics costs.

	Road transport	**Multimodal transport rail/road frequency (number of departures per week)**						
		7	**6**	**5**	**4**	**3**	**2**	**1**
Transportation costs	420.00	310.00	310.00	310.00	310.00	310.00	310.00	310.00
Cost of cycle stock	34.34	34.34	34.34	34.34	34.34	34.34	34.34	34.34
Cost of inventory in transit	334.25	358.90	361.19	364.38	369.18	377.17	393.15	441.10
Cost of safety stock	194.35	202.79	203.79	205.30	207.81	212.61	224.34	272.79
Total logistics costs	982.94	906.03	909.32	914.02	921.33	934.12	961.83	1058.23
	road transport	Multimodal transport barge/road frequency (number of sailings per week)						
		7	6	5	4	3	2	1
Transportation costs	42.00	270.00	270.00	270.00	270.00	270.00	270.00	270.00
Cost of cycle stock	34.34	34.34	34.34	34.34	34.34	34.34	34.34	34.34
Cost of inventory in transit	334.25	372.60	374.89	378.08	382.88	390.87	406.85	454.79
Cost of safety stock	194.35	205.92	206.91	208.39	210.87	215.60	227.17	275.13
Total logistics costs	982.94	882.87	886.13	890.82	898.08	910.81	938.36	1034.26

If one can assume demand during lead time to be normally distributed, the value of k that corresponds to a probability mass of α in the right tail of the distribution can be easily obtained from a table of the standard normal distribution.

For the case at hand, the (very high) cycle service level of 99% implies a k factor of 2.33 and leads to the following calculations.

Table 8.8 illustrates the importance of the frequency of multimodal transport in order to be competitive with direct road transport. At the time of the case, there were 5 rail connections per week and 3 sailings by barge available from Antwerp to Duisburg, suggesting multimodal barge/road transport at 910.81 EUR to be the cheapest option when comparing the cost of sourcing a single container.

8.5 Reconsidering the Case

Although the calculations from Section 8.3 offer a convenient way to estimate total logistics costs of sourcing a single container, some of the assumptions made can be easily violated in practice.

First, although the cycle service level is widely used in practice, in Dullaert *et al.* (2007) it has been shown that it tends to favor transport modes shipping smaller quantities as it does not take into account the number of replenishments needed per year. Instead, the fill rate that measures the fraction of demand that can be satisfied from stock should be used.

For a fill rate of *fr* per cent in the case of complete backordering and a normally distributed demand during lead time, one should select the safety factor k for which the following equality holds (Silver, Pyke, and Peterson, 1998, p. 268):

$$\int_{k}^{\infty} (u_0 - K)\frac{1}{\sqrt{2\pi}}\exp(-u_0^2/2)\mathrm{d}u_0 = \frac{Q}{\sigma_{\mathrm{DDLT}}}(1 - fr) \tag{8.3}$$

where Q = shipment size, σ_{DDLT} = standard deviation of demand during lead time. The right-hand side of Eq. (8.3) can be easily calculated and allows one to find the corresponding value of k from dedicated tables that can a.o. be found in Silver, Pyke, and Peterson (1998).

Equation (8.3) clearly shows that, contrary to the cycle service level, the fill rate will generate different safety factors for transport modes with a different loading capacity Q. In general, larger (smaller) shipment sizes increase (decrease) the right-hand side of Eq. (8.3), which in turn leads to lower (higher) values for the safety factor k to compensate for the lower (higher) risk of a stockout associated with the number of replenishment cycles per year.

Although the minimum quantity to be shipped is a single container, the shipment quantity sent by each of the three hinterland transport modes should not be identical as it is determined by the annual demand of 182 containers per year, the sailing frequency between New York and Antwerp and each mode's loading capacity. If there were only a single sailing per week, the shipment quantity to the port of Antwerp should be on average 182/52 = 3.5 containers and sourcing via road would have to result in a higher safety factor than multimodal rail/road and barge/road for which the oncarriage is assumed to have a fixed lead time.

Although the way in which the service level is expressed is likely to be an issue in multimodal transport, the main problem is probably that the lead time (and as a result the demand during lead-time distribution) often cannot be assumed to be normally distributed. The assumption of normality is, however, often made to facilitate the calculation of the safety factor for the cycle service level and the fill rate.

Whereas the minimum transport time of any leg in a supply chain is bounded by the physical maximum speed it can achieve and most often does not deviate that much from the most common duration, the maximum duration can be a lot larger than the most common transportation time. Indeed, strikes in ports, on (hinterland) terminals, severe congestion due to accidents or severe weather conditions can lead to transportation times that are a lot longer than the most common duration, albeit with a small probability. As a result, transport times for individual legs are usually positively skewed. Non-normal demand during lead time is an active research topic in the logistics literature (see e.g., Vernimmen *et al.*, 2008 for an extensive literature review) but to the best of our knowledge no specific attention has been paid to the implications of multimodal transport.

When adding up different legs in the transportation chain, one could expect the overall lead time to approximate the normal distribution, thanks to the central limit theorem. Because hinterland rail and barge services from a port have scheduled departure times servicing various sea-going vessels in the port, their departure cannot be delayed when a vessel's arrival would be delayed. As such, the delay in one leg of the supply chain can create additional delays by increased waiting times until the next scheduled transport of the next leg in the transport chain. The lead-time distribution of the individual legs are therefore not independent, preventing the central limit theorem from applying and resulting in (severely) positively skewed distributions even if the underlying distribution of the constituting legs were already normally distributed.

For the ease of exposition, consider a supply chain departing from New York to Antwerp 7 times a week and offering 3 departures to Duisburg (Monday, Wednesday and Friday, respectively) and that the plant operates 7/7. Upon arrival, containers can be moved instantaneously to the barge if available. Container transport from the port of New York takes on average 12 days with a variance of 1 day^2. Assume that the barge for example, departs on day 14 (Friday) with an average transport time of 1 day and a variance of 0.1 day^2. Ships arriving later than day 14, say at 14.02 days will have to wait until the departure scheduled at day 16 (Monday). Even if both the transportation time of the sea and the barge transport were normally distributed, a simulation on 10 000 ship arrivals and barge departures clearly shows that the resulting transport lead time (including the 0.5 day on carriage from the terminal in Duisburg to the site in Dortmund) show that when there are two days between consecutive departures, the overall lead-time distribution is clearly not normally distributed (Figure 8.3). Random numbers are generated by means of the leading Mersenne Twister random number generator (Matsumoto and Nishimura, 1998) and the variance reduction technique of common random numbers is used to be able to compare the different levels of lead-time variability under similar experimental conditions.

Table 8.9 shows the safety stock in number of container loads based on the simulation approach by considering the percentile associated with each service level and deducting the average demand during lead time, a hybrid approach using the analytical approach presented in Section 8.3 for the average and lead-time variance obtained by the simulation procedure ($\bar{L} = 14.96$ and variance $\sigma_L^2 = 1.18$), and an analytical approach using the values from Table 8.4 for 3 departures per week ($\bar{L} = 14.27$ and variance $\sigma_L^2 = 1.55$).

The difference in safety stock between the simulation and hybrid approach are therefore only due to the shape of the demand during the lead-time distribution. Note that for the case at hand the difference between the average lead time and variance of lead time remain small for the hybrid and analytical approach, and how the safety stock suggested is only 66 to 12% of the amount suggested by the hybrid or analytical approaches for cycle service levels of 65% or higher. For lower service levels, safety stock is even negative, a phenomenon first observed by Chopra, Reinhardt, and Dada (2004) and explored for its impact on the value of speed and reliability in Dullaert and Zamparini (2011).

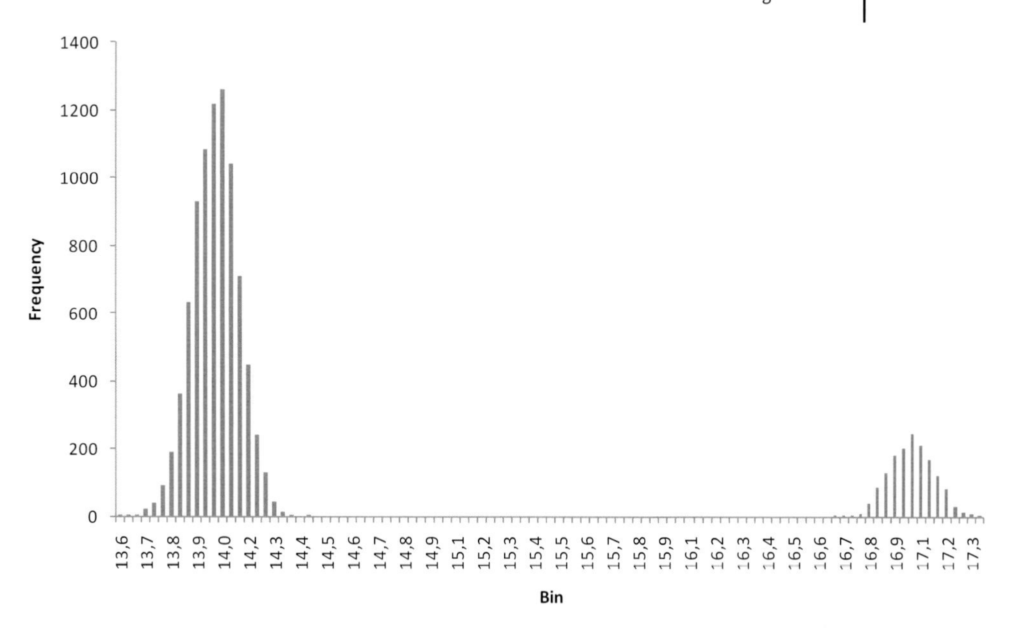

Figure 8.3 Demand during lead-time distribution the same goes for the demand during lead-time distribution depicted in Figure 8.4.

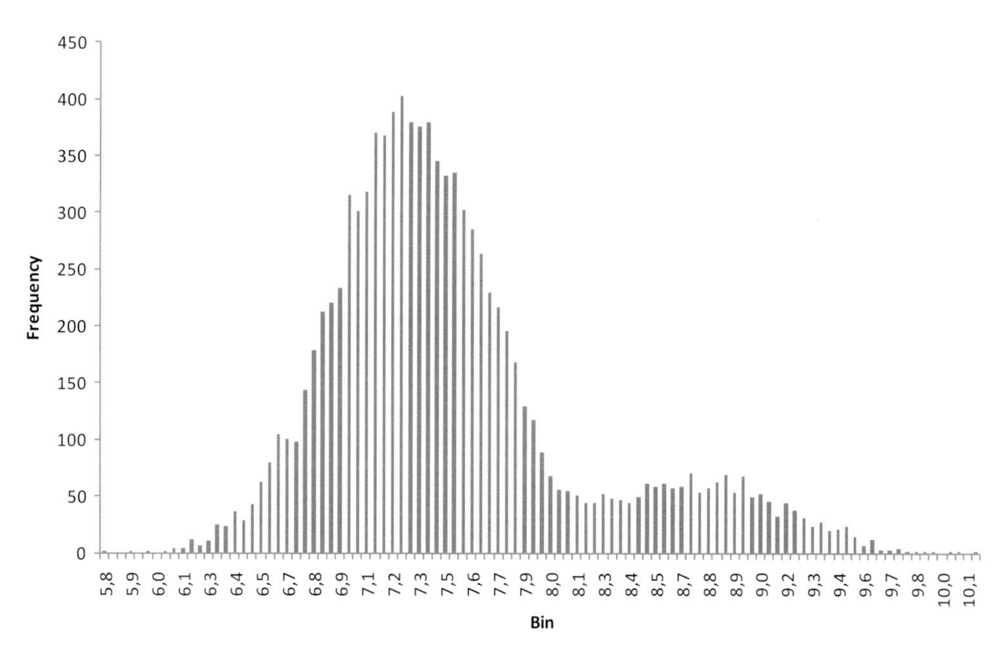

Figure 8.4 Demand during lead-time distribution show the resulting demand during lead-time distribution assuming that demand is normally distributed.

Table 8.9 Comparison of simulation and analytical results.

Service level	SS_simulation	SS_hybrid	SS-analytical
0.990	1.903	3.113	3.134
0.950	1.459	2.201	2.216
0.900	1.093	1.715	1.726
0.850	0.671	1.387	1.396
0.800	0.371	1.126	1.134
0.750	0.241	0.903	0.909
0.700	0.144	0.702	0.706
0.650	0.065	0.516	0.519
0.600	−0.003	0.339	0.341
0.550	−0.069	0.168	0.169
0.500	−0.129	0.000	0.000

8.6 Conclusions

The statistics related to the three considered geographic regions (EU, US and ASEAN) have shown that in the last decades there has been a remarkable increase in the goods shipped through multimodal transport and that the next decades will witness further increases in this segment of the transport market. Moreover, regions that are lagging behind in this respect are striving to invest in the infrastructures and in the standards that are needed to improve the efficiency of multimodal transport. In this respect, the combination of several transport modes to move cargo from an origin to the final destination, forces decision makers to choose between alternative multimodal transport options. In order to make economically sound choices, the assessment of the overall lead time and of the variance of lead time of the entire chain is needed.

The second part of this chapter has thus presented a basic framework to support such an analysis and has tried to draw attention to two model features that can have a significant impact in practice: the measurement of the service level and the shape of the demand during lead-time distribution. Depending on the actual case at hand, the basic frame can either lead to significant over- or underestimations of the required safety stock and hence of the overall total logistics cost of multimodal and unimodal transport alternatives. As a general rule, we therefore recommend the use of fill rates and of simulation modeling to support multimodal decision making in practice.

Bibliography

Baumol, W.J. and Vinod, H.D. (1970) An inventory theoretic model of freight transport demand. *Management Science*, **16** (7), blz. 413–421.

Bhattacharyay, B.N. (2010) Infrastructure for ASEAN connectivity and integration. *ASEAN Economic Bulletin*, **27** (2), 200–220.

Blauwens, G., Vernimmen, B., and Witlox, F. (2003) De generiek van het "total cost concept" [in Dutch], in *Transportgids*, Vol. 3.3.2/1-3.3.2/33 (ed. E. Claessens), Kluwer Editorial, Diegem, pp. 15–47.

Chopra, S., Reinhardt, G., and Dada, M. (2004) The effect of lead time uncertainty on safety stocks. *Decision Sciences*, **35** (1), 1–24.

Dullaert, W., Aghezzaf, E.-H., Raa, B., and Vernimmen, B. (2007) Revisiting service level measurement for an inventory system with different transport modes. *Transport Reviews*, **27** (3), 273–283.

Dullaert, W. and Zamparini, L. (2011) *Assessing the impact of reliability in freight transport: a transdisciplinary approach*, working paper, 19 p.

European Union (2010) *Energy and Transport in Figures 2010: Part 3 Transport*, 98 p.

Faust, P. (1985) *Multimodal Transport, Port Management Textbook – Containerization*, Institute of Shipping Economics and Logistics, Bremen, pp. 219–231.

Goh, M.K.H., DeSouza, R., Garg, M., Gupta, S., and Lei, L. (2008) Multimodal transport. A framework for analysis, in *Advances in Industrial Engineering and Operations Research* (eds A.H.S. Chan and S. Ao), Springer, New York, pp. 197–208.

Hayuth, Y. (1987) *Intermodality: Concept and Practice*, Lloyds of London Press Ltd, London, 149 p.

Matsumoto, M. and Nishimura, T. (1998) Mersenne Twister: A 623-dimensionally equidistributed uniform pseudo random number generator. *ACM Transactions on Modelling and Computer Simulations: Special Issue on Uniform Random Number Generation*, **8** (1), 3–30.

Silver, E.A., Pyke, D.F., and Peterson, R. (1998) *Inventory Management and Production Planning and Scheduling*, 3rd edn, John Wiley & Sons, Inc., New York.

UIRR (2011) *International Union of Road Rail transport Companies*. http://www.uirr.com/en/road-rail-ct.html (accessed 30 March 2011).

UN-ESCAP and AITD (2007) *Toward an Asian Integrated Transport Network, New York, UN.*

US Department of Transportation (2009) *Commodity Flow Survey 2007, Washington, US.*

US Department of Transportation (2010) *Freight Facts and Figures, Washington, US.*

Vernimmen, B. (2004) Een logistieke analyse van de modale keuze in het goederenvervoer [in Dutch]. G. Blauwens, P. d'Haese, and A. Van Breedam (eds.), *Logistiek. Laatste front in de concurrentieslag, 26th Vlaams Wetenschappelijk Economisch Congres, 25–26 March 2004, Antwerp*, 202–228.

Vernimmen, B., Dullaert, W., Willemé, P., and Witlox, F. (2008) Using the inventory-theoretic framework to determine cost-minimizing supply strategies in a stochastic setting. *International Journal of Production Economics*, **115** (1), 248–259.

9
Multimodal Analysis Framework for Hazmat Transports and Security

Cathy Macharis, Koen Van Raemdonck, Juha Hintsa, and Olivier Mairesse

9.1
Introduction

To conduct a well-balanced risk policy for hazmat transport, there is a need for a calculation method and analysis system. This analysis framework will keep the risk level acceptable in the case of new developments, detect and remediate existing bottlenecks and make it possible to clearly communicate about the risks. We make a clear classification between unintended incidents (in the case of accidents) and intended incidents (due to terrorist attacks, theft, sabotage and the like).

This chapter describes the risks of hazmat transport and how to deal with it. It shows the development of a methodology that will permit estimatation of the risks concerning the transport of dangerous materials in a structural way for unintended incidents. This should lead to an indication of the risk and consequences of a severe incident with dangerous goods along a specified route for each transport mode. In time, an estimation of the location of the vulnerabilities within the transport infrastructure network should be possible. For intended incidents, methodological issues are also discussed.

Many studies chose multicriteria analysis (MCA) or multicriteria routing models as a starting point regarding a calculation methodology for the estimation of the risk associated with hazmat transport. The possibilities concerning the estimation of the (in) security of a certain transport route, however, can be optimized by making use of the historical accident data. In the refined methodology it will be possible to draw a global risk map and a local risk map for the investigated area, and a methodology for the calculation of a local accident risk is offered, which takes local infrastructure parameters and accident data into account. This new evaluation framework will make it possible to estimate and compare the risk of hazmat transports by road, train, inland navigation and even pipelines. For intended incidents, this kind of probabilistic modeling is less applicable due to the scarcity of historical data. In this case, MCA methods show to be very useful as a means to select the most appropriate multimodal trajectory.

The first section consists of a literature review in which the different hazmat transport criteria, some existing risk analysis systems and security are being

Security Aspects of Uni- and Multimodal Hazmat Transportation Systems, First Edition. Edited by Genserik L.L. Reniers, Luca Zamparini.

discussed. Subsequently, a closer look is taken at the statistical analysis of accident data and the problems occurring with this analysis. Next, a new approach for calculating the local probability of the occurrence of a hazmat transport accident, taking into account local infrastructure parameters, is presented. Finally, an MCA model is developed for modal choice, taking security aspects into account.

9.2 Literature Review

9.2.1 Unintentional Incidents

Literature regarding the risk of hazmat transport is quite extensive. Usually multicriteria analysis or multicriteria routing models are being used as the starting point regarding the calculation of this risk. Panwhar, Pitt, and Anderson (2000) developed a risk analysis system for the transport of dangerous materials based on a geographical information system (GIS), which intended to reduce the impact of potential incidents involving hazardous substances. The risk for a hazmat accident to occur can be divided in two elements: (i) the probability of the occurrence of an accident and (ii) the consequences of an accident if it has occurred. The method described by Panwhar *et al.* uses a probabilistic risk assessment framework that takes into account the probability for accidents to occur on each road segment, based on the present infrastructure characteristics. Within this framework it is important to note that that decisions regarding hazmat transport concern different policymakers, and that different, sometimes conflicting, criteria should be reckoned with. In this study, they seek the route that scores best on some criteria, such as travel time, the presence of vulnerable locations (e.g., schools, hospitals, etc.) and the proximity to emergency services. Later, additional criteria will be added to the model. The framework searches for Pareto-optimal routes, so that trade-offs can be made between the different criteria. This means that the optimal route will not necessarily score best for all the criteria. These routes are visualized in a GIS.

Leonelli, Bonvicini, and Spadoni (2000) also argue that a methodology on the basis of multicriteria routing models can be used for the selection of the preferred route for hazmat transports. Indeed, if the minimization of the risk is the only criterion, routes that are more than twice as long as the quickest alternative could be chosen as the optimal path, but these are not feasible for financial reasons. This means that multiple criteria should be taken into account in the analysis. They also indicate that, in some cases, various objectives have been added after assigning weight factors to them. These weights reflect the relative importance of each criterion. This makes the optimal path or route dependent on the decision maker, who has to decide how to fix the value of the criteria weights. An example of such an MCA for hazmat transports is the TRANS methodology (Reniers *et al.*, 2009). In this study, the criteria were determined based on information out of the literature and expert knowledge. The values of the weights of these criteria were also mainly determined by the verbal method, namely on the basis of literature studies, assessments by experts in organized workshops and case studies. By adding the weights of the criteria for each segment, a score can be calculated per segment. Thereafter this score is multiplied with the score of a range, among which the number of transported dangerous substances on that road segment belongs. This final score is then plotted on the *Y*-axis of a chance–effect diagram and represents the probability of occurrence. It has to be noted that a potential issue with such an approach is the valid allocation of the weights, as there may be different optimal routes depending on how these values or weights are determined.

Lepofsky and Abkowitz (1993) conducted a similar multicriteria analysis. They demonstrated that the impact of hazmat transports can be calculated more effectively by using methodologies based on GIS models. The authors illustrated these methods by applying them in several case-studies in California. Huang, Cheu, and Liew (2004) explored a new approach for evaluating hazmat transports by using geographic information systems (GIS) and genetic algorithms (GA's). GIS is used to quantify the factors on each segment that contribute to the assessment criteria for a possible route, while GAs are applied to determine the weights of the different factors efficiently. In this way, the generalized costs of the alternative routes can be calculated.

Similar studies about the criteria of hazmat transports can, among others, be found in Abkowitz, Lepofsky, and Cheng (1991), in which the impact of the use of alternative criteria for route selection and the weights of these criteria are investigated; Abkowitz and Cheng (1988), who developed a methodology that takes risk and costs into account when searching for the optimal path for the transport of dangerous goods; and Saccomanno and Chan (1985) who discuss three routing strategies for the road transport of hazardous materials: (i) the minimal risk strategy, (i) the lowest accident risk strategy and (iii) the minimal operating costs strategy.

Clark and Bakersfield-Sacre (2009) argue that, while in the previous studies the risk of a particular route was analyzed starting from methods based on operations research (multicriteria analysis or multicriteria routing models), there is only little or no focus on probabilistic or statistical approaches, in which historical accident data can be used. Lord and Mannering (2010) give a detailed overview of the statistical models that can be used to analyze accident data, each with their advantages and disadvantages. The most commonly used of these methods are shown in Table 9.1.[1]

It can be concluded that MCA is a useful and quick screening tool to determine the safety of hazmat transport routes. It is, however, recommended to use quantitative risk analysis (QRA) for a more accurate and profound analysis and to avoid the risk for double counting and rank reversal and limit possible invalid weights (United Nations Economic and Social Council, 2008; Van Geirt and Nuyts, 2006).

1) See Lord and Mannering (2010) for a more complete overview.

Table 9.1 Advantages and disadvantages of the most commonly used statistical accident data models.

Model	Advantages	Disadvantages
Poisson	Basic model; easy to estimate.	Does not account for over- and underdispersion; negatively influenced by low sample mean and small sample size bias.
Negative binomial/ Poisson-gamma	Easy to estimate; able to handle overdispersion.	Does not account for underdispersion; can be negatively influenced by low sample mean and small sample size bias.
Zero-inflated Poisson and negative binomial	Able to handle datasets with a large number of zero-crash observations.	May create theoretical inconsistencies; zero-inflated negative binomial can be negatively influenced by low sample mean and small sample size bias.
Negative multinomial	Does account for overdispersion and serial correlation; panel count data.	Cannot handle under-dispersion; can be negatively influenced by low sample mean and small sample size bias.

Source: Lord and Mannering 2010.

The lessons learned from the literature review eventually argue in favor of the refinement of the existing methods by connecting them to accident data. Later in this chapter, a refined approach is introduced in which historical accident data and local infrastructure parameters are used to calculate the risk of hazmat transport in Flanders.

9.2.2
Intentional Incidents

Next to unintentional accidents, hazmats (or dangerous goods), are subject to a variety of man-made, illicit acts in the supply chain. Little is reported about how to deal with these incidents. Hintsa (2011), in his proposition for a "supply chain crime taxonomy", identifies over 20 different crime types, of which the majority may apply in hazmat supply chains, varying from theft to customs-law violations, from sabotage to document forgery, and from terrorism to violation of transport safety regulations, just to name a few examples (Hintsa, 2011). Hesketh (2010) specifically notes the risks with undeclared dangerous goods in the containerized maritime transport, using observations from MSC Napoli,[2] and the cargo which was rescued (dry) from the sunken ship for investigation pur-

2) See more information for example, at: http://www.bbc.co.uk/devon/content/articles/2007/07/16/napoli_timeline_feature.shtml (Oct.2008).

poses: "... dangerous goods are thought to account for between 5 to 10 per cent of total containerized cargoes but it is clear that some are carried undeclared ...". Regarding hazmats and acts of terrorism, Sheffi (2001) calls for increased attention on the movements of hazmat, soon after the 9/11 terrorist attacks: "... The threat of terrorism calls for further control of the movements of hazardous materials so that the authorities can react after a trailer load or rail car loaded with hazardous materials is reported missing, but before it is used in a terrorist attack ...". By using a multicriteria theory-based methodology, Tsamboulas and Moraiti (2008) provide a ranking of potential target locations for terrorist attacks, these may be cities or areas within a city that could be expressed in a graphical form on a map. One *modus operandi* for such attacks is naturally hijacking of hazmat cargo anywhere in the transport system, and delivery of it to such a target destination.[3]

From a governmental perspective, GAO (2009) is concerned about a too narrow risk-management approach, when it comes to freight rail and hazmat: "... Transportation Security Administration, TSA's efforts to assess vulnerabilities and potential consequences to freight rail have focused almost exclusively on rail shipments of certain highly toxic materials, in part, because of concerns about their security in transit and limited resources."

For the purpose of this chapter, the approach of "all potential crime types in supply chains" (Hintsa, 2011) is narrowed down towards the description of "hazmat security" by the International Road Union, IRU:

"Security refers to the measures and precautions taken to minimize the risk of theft or abuse of dangerous goods through which persons or the environment might be endangered." (IRU, 2005).

9.3 Refined Approach for the Calculation of Multimodal Hazmat-Transport Risk

In this methodology, a global risk map will be drawn, which is based on an average probability of occurrence of a catastrophic incident on a segment of a fixed length. This can be the catastrophic rupture of the truck tanks, rail wagons, and pipelines, and a large leak for inland vessels. Next, the risk is calculated for each segment as the product of probability and consequence. But first, some general principles of the methodology are introduced.

9.3.1 Preliminary

In this approach for the estimation of the risk of multimodal hazmat transports, a general assumption is that the hazardous substances are classified into four types based on the following scenarios:

3) This research was carried out within the European COUNTERACT-project.

- puddle fire for *flammable liquids*;
- evaporating puddle for *toxic liquids*;
- toxic cloud for *toxic gases*;
- BLEVE ("boiling liquid expanding vapor explosion") for *flammable (liquefied) gases*.

Another important assumption is that the risk of a catastrophic incident can be divided into two distinct parts that can be validated on the basis of historical incident and/or crash data.

1) the general probability calculation based on international incident data of transport of hazardous substances – this is the basis for the global risk map;

2) local probability calculations based on accident data and infrastructure parameters of the complete available freight transport in Flanders – this is the basis for the local risk map.

These 2 components are connected through the following relationship:

$$\begin{aligned}&\textit{Probability of occurence of a catastrophic incident}\\&\quad = \textit{General probability } P \times \textit{Locality parameter } C\end{aligned}$$

The locality parameter (C) is a coefficient that reflects the location specific circumstances that may lead to an incident with any freight. The first thing that has to be done in the refined approach is to divide the trajectory into different segments with a fixed length. Next, a global risk map is set up, based on the general chance of occurrence of a catastrophic incident as shown above, and the consequences of such an incident. If the local infrastructure parameters of the segments are known, a local risk map can be developed in more detail. This methodology can be applied analogously for the different transport modes. The result of these calculations can then be visualized on a geographical map.

9.3.2
Step 1: Determination of the Segmentation and Estimation of the Impact Distances

As already mentioned, the first step of the methodology is to divide the transport trajectory into different segments, which will make it possible to determine, among others, the consequences of a catastrophic incident involving hazmat transports. These consequences are expressed as a number of potential victims within a certain impact distance.

To determine these consequences, the trajectory has to be divided into segments of a length L and around each segment an impact distance R is plotted from the center of that segment. The smaller the segment length, the more accurate the results of the calculation of the potential victims. In this case, a length of 50 m was chosen, because this is also the length of the smallest impact distance. The segment length may never exceed two times the impact distance, because the

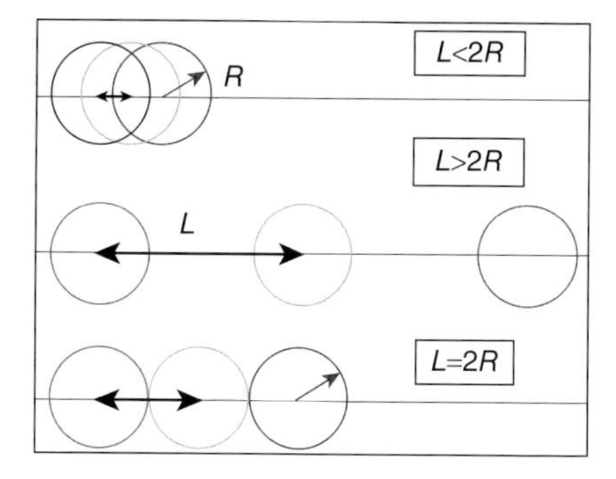

Figure 9.1 Segment length versus impact distance. Source: LNE, 2010.

Table 9.2 Impact distances for the different types of hazardous substances and transport modes.

Type of hazardous material	Impact distance "Road"	Impact distance "Railway"	Impact distance "Inland navigation"
Flammable liquid	50 m	50 m	150 m
Toxic liquid	200 m	150 m	500 m
Flammable (liquefied) gas	350 m	475 m	225 m
Toxic gas	375 m	625 m	725 m

Source: LNE 2009.

number of potential victims is calculated within an impact zone corresponding to a circle with center in the middle of the segment and a radius equal to the impact distance. If the segment length is greater than twice the impact distance, potential victims present in areas outside the circle will not be included in the calculations (as can be seen in Figure 9.1).

The impact distances were classified for four types of hazardous substances (flammable liquids, flammable gasses, toxic liquids and toxic gases) and were determined in a study of LNE (2009). The impact distances for each type of hazardous substance and transport mode can be found in Table 9.2.

The potential victims are those that are present within the impact distance at the moment of the incident and should include the residents, the individuals present in vulnerable locations (i.e. schools, hospitals and nursing homes), individuals present in companies, industrial areas, public places, and road users. Based on a data layer "statistical sectors", in which data about the number of residents, hospitals, etc., are available, potential victims can be calculated. To do so, the statistical sectors with which the impact area overlaps are selected. Thereafter, the area of the overlap is expressed as the percentage of the total size of the corresponding statistical sector. The number of victims that should be assigned to the impact zones can be obtained by adding up the products of the calculated percentages and the population densities of the corresponding sectors.

9.3.3
Step 2: Estimation of the General Probability of Occurrence of a Catastrophic Incident

The general probability of the occurrence of a catastrophic incident consists of the probability of release on the one hand, and the conditional probability of subsequent events, once the release has occurred on the other hand:

$$\begin{aligned} &\textit{Probability of an incident with catastrophic result} \\ &\quad = \textit{Probability of release} \times \textit{Conditional probability of a subsequent event} \end{aligned}$$

In particular, the general probability of a catastrophic hazmat incident on a specified route for each year is determined by the following formula:

$$P_{\mathrm{cithm}} = F_{\mathrm{ci}} \times Q_{\mathrm{thm}}$$

With:

P_{cithm} = Probability of the occurrence of a catastrophic incident involving a certain type of hazardous material;
F_{ci} = Frequency of a catastrophic hazmat incident (expressed in tkm);[4)]
Q_{thm} = The quantity of a specific type of hazardous substance on a specific route (expressed in tkm/year).

9.3.4
Step 3: Estimation of the Local Probability of Occurrence

As explained above, the general probability of the occurrence of a catastrophic incident with hazmat transport is calculated by multiplying the number of tonnes of one type of hazardous substance being transported on a given route with the frequency of hazmat incidents per tkm. However, transportation incidents are also influenced by local risk factors included in the traffic system such as infrastructure and traffic parameters. The examination of the local transport conditions that lead to traffic incidents compared to the average transport conditions in the whole region allows us to introduce a factor with which the general probability of occurrence of a catastrophic incident can be multiplied, in order to make a differentiation in function of the local circumstances. These local circumstances were identified and translated to a number of infrastructure parameters in the TRANS methodology (Reniers *et al.*, 2009). The intention of this locality factor is to evaluate the risk of different routes, that is, where the probability of an accident is greater or smaller than the average probability for freight transport, due to the local infrastructure parameters.

4) tkm = ton-kilometer.

The locality parameter C can be estimated by dividing the actual number of accidents for a certain segment of a given length by the total number of accidents on the entire transport network, reduced to the average on same segment length. This is shown in the following formula:[5)]

$$C = \frac{y_{\text{loc}}}{y_{\text{avg}}}$$

In which:

$$y_{\text{avg}} = y_{\text{total}} \times \left(\frac{L}{L_{\text{total}}} \right)$$

and:

y_{loc} = number of accidents on segment of length L;
y_{avg} = average number of accidents per segment of length L;
y_{total} = total number of accidents on the transport network;
L_{loc} = total length of the examined transport network;
L = length of the segment for which the locality parameter is calculated.

On each segment of an existing route a locality parameter can be attributed. This exercise can be done for existing roads or routes and allows us to globally take into account the local characteristics of the route. The influence of a specific infrastructure parameter, however, cannot be deduced by using this approach. To estimate the influence of the specific infrastructure parameters on the total accident rate, and to estimate the locality parameter for a possible new route, a detailed statistical study has to be performed.

The influence of the infrastructure parameters can be deduced from the historical accident data, and used as input into a predictive model. The weights (regression coefficients) obtained out of the model give an indication of the relative importance of the various infrastructure parameters, and can then be used to calculate a local probability of the occurrence of an accident on the basis of the presence of the local infrastructure parameters on that segment. The ratio between the local probability of occurrence and the average probability of occurrence gives the locality parameter C_{loc}.

Although empirical research shows that accident data approach Poisson-like distributions, it is important to always check the distribution of the dependent variable before determining which generalized linear model will be applied. Indeed, the distribution of accident data is likely to vary according to the width of the selected time frame or the length of the segments. To simplify the explanation of the calculation methodology of the locality parameter, it will be assumed that the time frame and the segment length are chosen so that the accident data are normally distributed. The theoretic explanation of the statistical models that can

5) Notice that the locality parameter is calculated for accidents, that is, unintentional incidents specifically, because it is directly related to databases of unintentional incidents. However, the former part of the general equation is applicable to both types of incidents.

be used to determine the locality parameter is not described in detail, because multivariate analysis is beyond the scope of this book.[6)]

Assuming that the accident data are normally distributed, the calculation of the locality parameter C_{loc} can be formulated as follows:

$$C_{loc} = \frac{y_{loc}}{y_{avg}} = \frac{\left(b_0 + \sum_{i=1}^{n} b_i x_{i_{loc}}\right)}{\left(b_0 + \sum_{i=1}^{n} b_i x_{i_{avg}}\right)}$$

With:

b_0 = intercept;
b_i = regression coefficient for the infrastructure parameters (calculations based on the complete dataset);
$x_{i_{avg}}$ = average value of the infrastructure parameter per length L;
$x_{i_{loc}}$ = value of the infrastructure parameter per length L of the segment to be investigated;

If the normality assumption does not apply, one can opt for a negative binomial model, in which the variance does not have to be equal to the mean. In case of a large amount of zero-crash observations, the model can be estimated more correctly by using a zero-inflated Poisson-model instead of a regular Poisson-distribution. If data is only available when an accident actually happens, and thus $P(0) = 0$, it can be better to use a zero-truncated Poisson-model (Zuur *et al.*, 2009). Other possibilities are for example, multinomial logit models in which the link function is also a logit function. For exponential, gamma or inverse Gaussian models, inverse or squared inverse functions are used as the link function (McCullagh and Nelder, 1989). In the analyses of accident data however, these models are not that common. For a more complete overview of the possible models we refer again to Lord and Mannering (2010).

9.3.5
Practical Application and Visualization

Since the local and general probability of the occurrence of a catastrophic accident are both known, and it is known how to determine the consequences of a catastrophic incident, the risk can be calculated as well. The *global risk map* is a visual representation of the external human risk on a geographical map, based on historical accidents. This global risk map consists of a set of maps (13 per transport mode) that visually captures the risk of hazmat transports for each transport route. Per type of hazardous substance a *consequence map*, which visualizes the number of potential victims along the route, and a *probability map*, which visualizes the probability of the occurrence of a catastrophic

6) More information on these statistical models can be found in Garson (2010), Van Geirt and Nuyts (2006), JTP (2010), Lord and Mannering (2010), and Lord *et al.*, (2004).

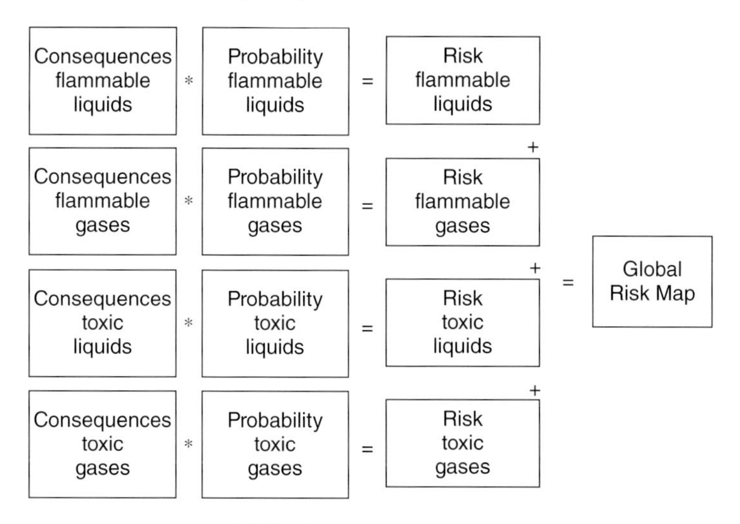

Figure 9.2 Setting up global risk map. Source: LNE 2010.

accident along the route, is set up. With these two maps it is possible to set up a *risk map* per type of hazardous substance. By adding up the 4 risk maps the *global risk map* can be found. This procedure can be followed for each transport mode. How these different maps are connected to each other is shown in Figure 9.2.

For each segment, the risk is defined by multiplying the probability and consequence. The segment risks are visualized by a risk map for each transport mode and each type of hazardous substance. The total risk, shown in the global risk map, is obtained by adding up the four risk maps of the four types of hazardous substance. An example of a global risk map that was obtained in this way can be seen in Figure 9.3.

9.3.6
Interpretation of the Results

The results allow comparing the different transport modes based on the average frequency of release and the different risk parameters, as well as the calculation of the number of potential victims along a certain route. The global risk maps provide an insight in the overall risk due to the transport of dangerous goods. The underlying risk maps for each of the four types of hazardous substances show the individual contributions of each type to the overall risk, for each transport mode. The methodology also allows estimation of the consequences of a possible increase or decrease of the transported volume on a route, or the effect of an increase or decrease in the number of people exposed along the route. The local probabilities

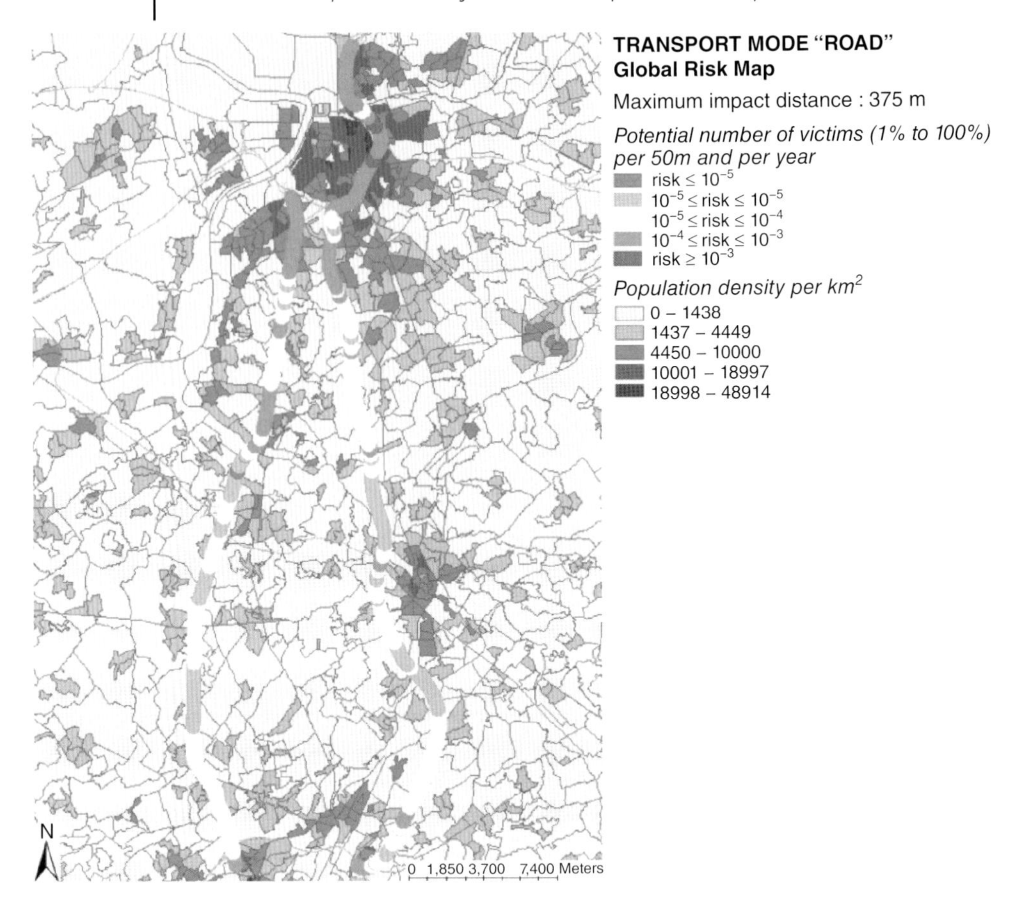

Figure 9.3 Global risk map for the transport mode "Road". Source: LNE 2010.

of occurrence of an incident can also show the influence of a change in the infrastructure parameters.

The risk maps can also support decision making. The methodology allows, for example, calculation of the risk of locating a new school or hospital close to a hazmat route. If the school or hospital happens to be located within the color band, it should be investigated whether the risk of this location can be accepted. For an existing vulnerable location, the risk map shows what risk people might be exposed to, for which specific plans should be developed to mitigate these risks. The underlying probability and consequence maps also allow investigating the different contributions to the risk on a specific location, namely a large number of people exposed or a high "locality parameter", or a large quantity of hazardous substances transported along that part of the route. One can also see what type of hazmat transport leads to the highest risk.

Table 9.3 Actors and locations of intended hazmat incidents.

Actors/explosion location	On-site (Where the incident first happens)	"Attractive target" (Outside of the normal route)	Random location (Either moving or standing)
Terrorists	Yes (Deliberately or accidentally)	Yes (Assuming hijacking of the load)	Yes (Deliberatelyor accidentally)
Cargo thieves	If there is an accident	No	If there is an accident
Activists	Yes (Deliberatelyor accidentally)	No	Yes (Deliberatelyor accidentally)

9.4 Intended Incidents with Hazmat Transport

Table 9.3 was created to illustrate the categories of illicit actors causing an incident in a hazmat supply chain, as well as to characterize the locations where the incidents could potentially take place.

First, looking at three types of illicit actors and the typical motivations they have (table below, left column):

- terrorists, who typically seek high damage rates (human casualties, property);
- cargo thieves, who seek high profits for stolen goods/materials (their interest is not to have the cargo or vehicles destroyed);
- activists, who seek publicity while protesting against specific industry sectors or companies (this interest would not be in causing damage to the surroundings).

Secondly, looking at where possible explosions could take place, following the intrusion by the illicit actor(s) (Table 9.3, first row):

- on-site (where the intrusion first happens, normally a place where the vehicle is at stand still);
- "attractive target", landmarks, dense population etc. (which would not be next to the normal transport route, due to safety regulations);
- random location (vehicle and cargo either moving or standing);
- looking finally at the three alternative scenarios where a hazmat explosion could take place, following an intrusion by an illicit actor:
  - On-site explosion happens with terrorists and activists, when they deliberately trigger the explosion immediately after their intrusion, or when there

is an immediate accident; with cargo thieves, an immediate explosion could happen by accident only.

- Explosion next to an attractive target could happen with the terrorism scenario, where the hijacked cargo is driven to the target destination, and an explosion is triggered there, to cause "maximum damage"; this normally would not happen with cargo thieves or activists.

- Explosion at random location (on road or at stand still) would mainly happen if there is an accident following a terrorist, cargo thief or activist action; with terrorists and activists, explosion at random location could also be intentional.

While the methodology described above, which was set up to determine segments and major axes on the transport network which are the most susceptible to accidents involving hazardous materials, was based on a probabilistic approach, it might still be more interesting to determine dangerous locations in terms of intentional incidents such as terrorist attacks, thefts, etc., using multicriteria analysis (MCA). Such an approach was developed by Tsamboulas and Moraiti (2008), who propose an evaluation methodology based on MCA to assess the degree of attractiveness of specific locations to terrorist attacks associated with freight transport.

Since the complete protection of the transport network is unrealistic and economically unfeasible, it is important to identify the locations that are the most susceptible to intentional incidents. Therefore, Tsamboulas and Moraiti present a methodology that takes into account different criteria to identify potential locations that could be risk prone to intentional incidents and for which security measures could be taken in times of heightened threat. The criteria are in fact the measures of performance by which any possible location will be assessed and prioritized. For each of these criteria a weight is determined, as well as a corresponding indicator. The values of these weights depend on the extent to which the criterion influences the attractiveness of a certain target. Finally, the overall degree of attractiveness of a location is calculated by adding the obtained weighted scores. Different locations can then be compared on the basis of this degree of attractiveness.

Tsamboulas and Moraiti selected five sets of criteria for assessing the attractiveness to intentional incidents related to freight transportation:

- public impact;
- economic impact;
- social and political impact;
- infrastructure;
- newsworthiness.

Of these criteria, public impact was found to be a very important criterion while identifying locations that are risk prone to intentional incidents with hazmat transports. This criterion reflects the impact on the life and well-being of the population exposed at a certain location, and also is a key factor in determining the consequences of an intentional incident. Indeed, these consequences can be calculated analogously to the potential victims within the affected area as explained in the methodology to identify dangerous segments on the transport network when looking at unintended incidents, presented above. Most similarities between the method to assess locations that are risk sensitive to intentional incidents with hazmat transports and the method that determines risk-prone sites for nonintentional incidents (e.g., traffic accidents) with hazmat transports can be found in the estimation of the consequences of the incidents.

9.5
How to Include Security in the Modal Choice

Another way to use multicriteria analysis in the field of security and multimodal hazmat transport is by using it as a tool to select the most appropriate transportation mode or combination of transportation modes. The modal choice can indeed be seen as a multicriteria decision. There are different modal choice variables that are important, such as transport cost, transport time, reliability, frequency, etc. (Shinghal and Fowkes, 2002; Vannieuwenhuyse, Gelders, and Pintelon, 2003; Norojono and Young, 2003). But also security is an important variable that is often taken into account by shippers and logistics providers (Tongzon, 1995). In what follows, we propose a multimodal choice tool, taking security into account.

There exist various techniques to conduct a MCDA. There is no better or worse technique, because the appropriateness of the method depends on the specific decision situation (Tsamboulas *et al.*, 1999). Among the methods, the most popular ones used in the field of transport are AHP, multiattribute theory variants (MAUT, MAVT, SMART, SMARTER, VISA), outranking methods (PROMETHEE, ELECTRE), regime analysis, UTA and TOPSIS (Macharis and Geudens, 2009).

9.5.1
Case Study

In this case study (based on Macharis, Heugens, and van Lier, 2008), a multicriteria analysis is used to determine the most interesting mode of transport for a specific trajectory of a shipper. In a multicriteria analysis, various alternatives are compared on the basis of several criteria and subsequently a global ranking of the alternatives is proposed (Belton and Steward, 2002). The trajectory goes from Flanders to Turkey, where three possible transport modes can be chosen: road, intermodal road/rail and intermodal road/short sea shipping. The goods (liquid cyanide) are packed in containers and the whole intermodal chain is included in the analysis. In the first section the multicriteria analysis method is explained. In the second section the various alternatives are described. The subsequent section presents the different modal choice variables and their importance to the shipper, while the variables are also compared to previous research. The alternative transport modes are assessed on the basis of the previously identified criteria in a fourth section, after which the actual multicriteria analysis can be performed. The used data are real and collected through a study of an international company. The evaluation of the security of the different transport modes was done through a small survey among shippers.

9.5.1.1 PROMETHEE

PROMETHEE is developed by Brans (1982) and further extended by Brans and Vincke (1985) and Macharis, Brans, and Mareschal (1998). It belongs to the methods of partial aggregation, also called outranking methods, and was partly designed as a reaction to the complete aggregation (MAUT) methods (De Brucker *et al.*, 2004). The evaluation table, where the alternatives are evaluated on the different criteria, is the starting point of the PROMETHEE method. The use of the PROMETHEE method requires additional information.

First, a specific preference function needs to be defined ($P_j(a,b)$) that translates the deviation between the evaluations of two alternatives (a and b) on a particular criterion (g_j) into a preference degree ranging from 0 to 1. This preference index is a nondecreasing function of the observed deviation (d) between the scores of the alternatives on the considered criterion ($f_j(a)–f_j(b)$), as shown in Eq. (9.1). In order to facilitate the selection of a specific preference function, six possible shapes of preference functions are proposed to the decision-maker by Brans *et al.* (1986) (usual shape, U-shape function, V-shape function, level function, linear function and Gaussian function).

$$P_j(a, b) = G_j\{f_j(a) - f_j(b)\} \tag{9.1}$$

Secondly, information on the relative importance of the criteria (weights, w_j) is required. PROMETHEE assumes that the decision maker is able to weigh the criteria appropriately, at least when the number of criteria is not too large (Macharis *et al.*, 2004). With this information, an overall preference index $\pi(a,b)$ can be computed, taking all the criteria into account (see Eq.(9.2)). This preference index is based on the positive $\varphi^+(a)$ and negative $\varphi^-(a)$ preference flows for each alternative, which measures how an alternative (a) is outranking (see Eq. (9.3)) or outranked (see Eq. (9.4)) by the other alternatives. The difference between these preference flows is represented as the net preference flow $\varphi(a)$ (see Eq. (9.5)), which is a value function whereby a higher value reflects a higher attractiveness of alternative (a).

$$\begin{cases} \pi(a, b) = \sum_{j=1}^{k} w_j P_j(a, b) & (9.2) \\ \varphi^+(a) = \frac{1}{n-1} \sum_b \pi(a, b) & (9.3) \\ \varphi^-(a) = \frac{1}{n-1} \sum_b \pi(b, a) & (9.4) \end{cases}$$

$$\varphi(a) = \varphi^+(a) - \varphi^-(a) \tag{9.5}$$

Three main PROMETHEE tools can be used to analyze the evaluation problem: (i) PROMETHEE I partial ranking, (ii) PROMETHEE II complete ranking and (iii) the GAIA plane. In PROMETHEE I, the partial ranking is obtained from the positive and negative outranking flows (see Eqs. (9.3) and (9.4)). In this respect, alternative (a) is preferred to alternative (b) if it has a high positive flow and a low negative flow. In some cases, the ranking of alternatives may be incomplete as PROMETHEE I allows indifference (both positive and negative flows are equal) and incomparability (alternative (a) scores high on a set of criteria on which (b]) is weak and *vice versa*) situations. PROMETHEE II provides a complete ranking of the alternatives from the best to the worst one, which is based on the net preference flow (see Eq. (9.5)). The geometrical analysis for interactive aid (GAIA) plane provides a graphical representation in which the alternatives and their contribution to the criteria are displayed. Additionally, a decision stick can be used to further investigate the sensitivity of the results in function of weight changes (Brans and Mareschal, 1994).

Recently, an extensive literature review on PROMETHEE methodologies and applications has been performed by Behzadian *et al.* (2010). It showed that PROMETHEE is increasingly used in a variety of domains such as environment management, logistics and transportation, energy management, and so on (see De Brucker *et al.* (2004), Kumar *et al.* (2006), Beynon and Wells (2008), Dagdeviren (2008), Tuzkaya (2009) and Mohamadabadi *et al.* (2009)).

9.5.1.2 **The Alternatives**

The case concerns the transport of a container load with a net volume of 70 m^3 from the production site near Oudenaarde to the destination in Istanbul, where several retail stores sell the products. A storehouse, located in Sefakoy – near Istanbul – is used as a local distribution center. Three transport options were considered: road transport, intermodal rail/road transport and intermodal short sea/road transport. Direct international transport via rail or short sea connections were excluded because the company/shipper did not have the required connection possibilities.

For the alternative with intermodal rail/road transport the rail terminal in Genk can be used in Flanders, while in Turkey, use can be made of the terminal in Halkali. This is visualized in Figure 9.4.

If the shipper decides to transport the goods by using short sea shipping (see Figure 9.5), the goods will be transported to Antwerp (95 km) by road first, where they will be transferred on a ship to Istanbul. The road from Istanbul to Sefakoy is 59 km. The ship leaves at one particular day in the week.

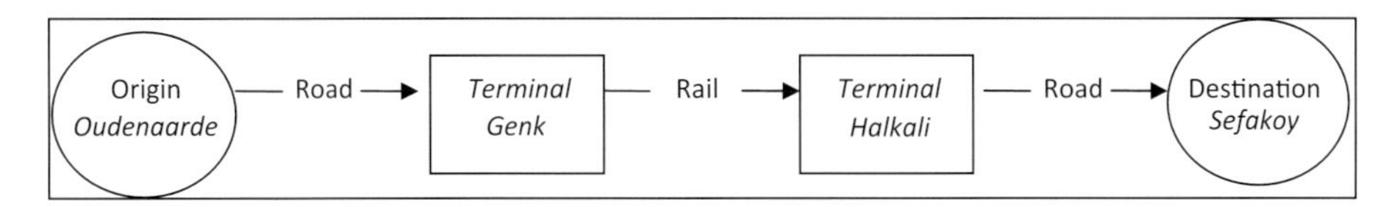

Figure 9.4 Intermodal transport chain: rail.

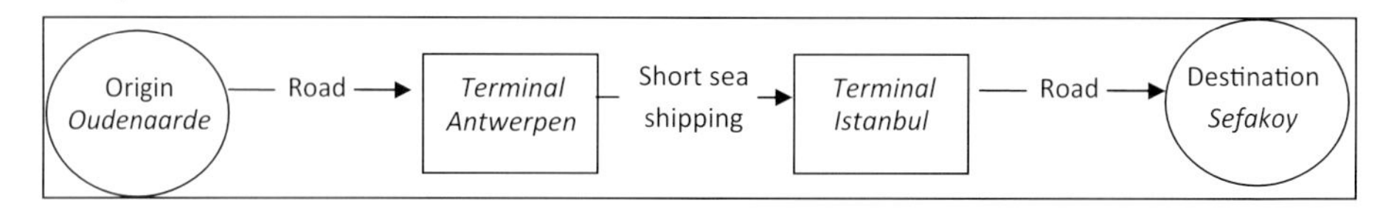

Figure 9.5 Intermodal transport chain: short sea shipping.

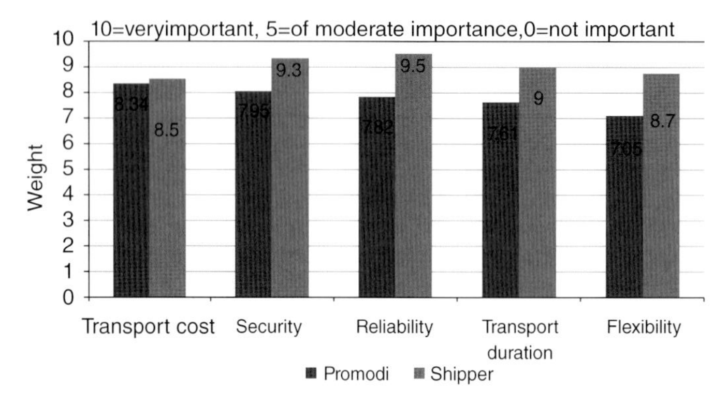

Figure 9.6 Comparison of the valuation between the Promodi respondents and the shipper.

9.5.1.3 **The Weights: Valuation of the Critical Factors Influencing the Choice of Transportation**

The main modal choice variables on which shippers base their transport decision were identified in the Promodi results report (Vannieuwenhuyse and Germis, 2002). For this purpose, the shippers were asked to weigh and rank the 11 Promodi factors on a scale from 0 to 10. Thus, 10 was very important, 5 was of moderate importance and 0 was not important at all. This *direct rating* methodology has been shown to be similar to empirically established weights, and being fairly easily obtainable (Zhu and Anderson, 1991). This study showed that transport costs, security, reliability, the duration of the transport and flexibility are the most important factors when making transport decisions. All these factors scored between 7.05 and 8.35 on a scale of 0 to 10, while the other variables scored significantly less.

Within this framework, the shipper of the investigated corporation was also asked to weigh and rank the 11 Promodi factors to find out which of these factors influence this shipper's transport decision the most. Like the majority of respondents in the Promodi study itself, the shipper in this case also attaches the highest value to transport costs, security, reliability, transport duration and flexibility. However, the variables were ranked in another order.

The order of importance of the variables according to the shipper in our case can be deduced from the graph in Figure 9.6.

1) Reliability;
2) Security;
3) transport duration;
4) flexibility;
5) transport costs.

The weights assigned to the different factors lie very close to each other, which implies that the shipper finds them all very important.

9.5.1.3.1 **Valuation of the Alternatives on the Criteria**

Total Logistic Cost Equation Based on the total logistic cost equation, the logistic costs for the three alternatives can be calculated. The used formula is the following, and was designed by Blauwens *et al.*, 2002).

Total logistic costs = transport costs + costs of cyclical inventory + inventory costs during transport + costs of safety inventory

Or in analytical form:

$$\mathrm{TLC} = \mathrm{TC} + \left(\frac{1}{R} \times \frac{Q}{2} \times w \times h_\mathrm{m}\right) + \left(L \times w \times \frac{h_\mathrm{v}}{365}\right) + \left(\frac{1}{R} \times w \times h_\mathrm{m} \times K \times \sqrt{(T \times d) + (D^2 \times t)}\right)$$

Commodity flow parameters:

R Annual volume that has to be transported (ton/year, containers/year . . .)
D Average daily demand or consumption from the inventory (ton, containers . . .)
d Variance of the daily demand/consumption from the inventory (ton, containers . . .)
w Value of the goods (€/ton, €/container . . .)
h_m Holding cost in the repository (%/year)
h_v Holding cost during the transport (%/year)
K Safety factor, depending on the permitted risk on inventory shortages during delivery time

Transport cost parameters:

TC Transport costs of the transport mode (€/ton, €/container . . .)
Q Load capacity of the transport mode (ton, containers . . .)
T Average delivery time of the transport mode (days)
t Variance of the delivery time of the transport mode (days^2)

The parameters in Tables 9.4 and 9.5 were used to calculate the total logistic costs:

Table 9.6 summarizes the transport cost calculations and allows concluding that intermodal freight transport to Sefakoy (Turkey) by rail and short sea shipping are, respectively, 6% and 47% cheaper compared to road transport.

Table 9.4 Commodity flow parameters for distribution to Turkey.

R	Yearly volume (m^3/year)	3.360
D	Average daily demand (m^3/day)	15
d	Variance of daily demand ($(m^3)^2$/day)	1.91
w	Value of the goods (€/m^3)	338
h_m	Holding costs in repository (% per year)	22%
h_v	Holding cost during transport (% per year)	20%
K	Safety factor	3.3

Table 9.5 Transport parameters for distribution to Turkey.

		Rail	Short sea	Road
TC	Transport cost (€/m^3)	?	?	?
Q	Load capacity (m^3)	70	70	70
T	Average delivery time (days)	5	10	5
t	Variance of delivery time (days^2)	2	4	1

Table 9.6 Difference in transport costs between intermodal transport and road transport.

	Intermodal transport by rail	Intermodal transport by short sea shipping	Road transport
Transport cost €/m^3	41.79	23.31	44.29
	94%	53%	100%
Difference with road transport	–6%	–47%	

The inventory costs are the highest for the slowest transport mode, being intermodal transport by short sea shipping. Figure 9.7 shows that the inventory costs for road transport and intermodal transport by rail count for, respectively, 6% and 7% of the total logistic costs. In the case of intermodal transport using short sea shipping this part is almost three times as high, that is, 17%. Further analysis shows that the high inventory costs can be entirely attributed to:

- *High inventory costs during transport* itself, since intermodal transport using short sea shipping is characterized by a slow delivery time: 10 days, compared to 5 days for road transport and intermodal transport by rail.
- *High safety inventory costs* because the variance of the delivery time is equal to 40% for intermodal transport (rail and short sea shipping) in comparison with 20% for road transport.

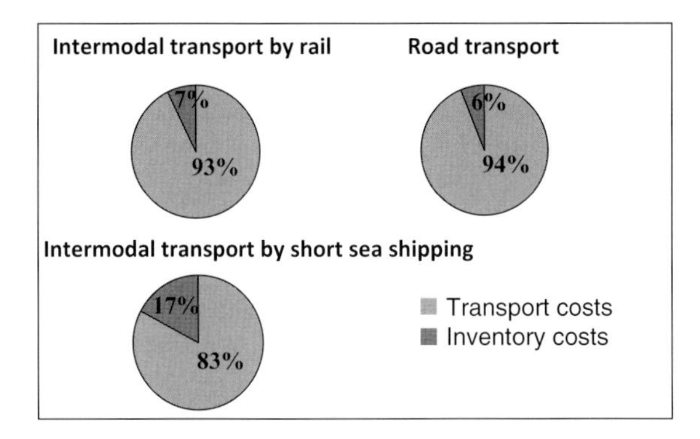

Figure 9.7 Transport costs versus inventory costs by mode (€/m³).

From a total logistic costs perspective, road transport is the most expensive alternative. Intermodal transport using short sea shipping and rail are 40% and 4% cheaper compared to road transport. However, the shipper will also take other modal choice variables into account, which are discussed more thoroughly in the following section.

9.5.1.3.2 **Evaluation of the Qualitative Variables**

Building on the findings from the previous section, in this section it will be examined how well or poorly the alternative intermodal transport modes score compared to each other and to road transport, from the shipper's point of view.

Security With respect to security, we take into account theft, sabotage and terrorism. Road transport is clearly the most vulnerable, when it comes to theft, sabotage and terrorism. According to European regulations, the drivers must take breaks based on the driving hours behind them, independently of whether secure parking lots are in the vicinity or not, thus making them extra vulnerable to such illicit acts. When it comes to sabotage and terrorism, these could take place also with moving trucks, for example, by targeting them with explosives. And, in the worst-case scenario for terrorism, a truck could be hijacked and driven into a highly attractive target, and destroyed there.

Rail can be considered to be less vulnerable than road in front of the illicit acts. Theft and sabotage with a moving train present big challenges for the criminal actors. Railyards and terminals remain the weakest link while trains are being loaded, or are stopped for other reasons. Hijacking and/or redirecting a train for terrorist purposes is of course possible, but not such a likely *modus operandi*.

Finally, short sea shipping can be seen as the least vulnerable of the three transport modes. Once goods have been loaded on board, it would require a serious internal conspiracy or act of sea piracy (or war-like attack on the ship) in order to carry out criminal and/or terrorist activities. Again, the ports of loading and unloading form the biggest vulnerabilities.

Flexibility Road transport scores the highest on flexibility since the shipper himself regulates the loading hours of the trucks and is not bound to fixed departure dates as is the case with rail or sea shipments. In addition, short sea shipping often offers fewer options concerning departure dates compared with rail transport, which therefore claims the second spot in the ranking. Reasons for this include:

- If a container is forgotten in the case of short sea shipping, it is possible that this container remains aside for a week, which results in a delayed delivery to the customer.
- Better transit times in the case of intermodal transport by rail.

Reliability Road transport is the most reliable transport mode from the shipper's point of view, but he has no preference between intermodal transport by rail or short sea shipping for this criterion. The reason for this is that short sea shipping has transit times that are a little longer, but rail transport is more sensitive to strikes.

It can be concluded that, according to the shipper, road transport scores better in terms of security, reliability and flexibility in comparison with intermodal transport by rail or short sea shipping.

9.5.1.4 Multicriteria Analysis

Based on the information above, an evaluation table can be set up (see Table 9.7). In this table the different alternatives (road transport, intermodal transport by rail and intermodal transport using short sea shipping) are being evaluated on the basis of the 4 modal choice variables.

For the fifth criterion security, we take into account 3 subcriteria (see Table 9.8):

Table 9.7 Evaluation table.

	Transport costs	Reliability	Transport duration	Flexibility
Road transport	44	5	5	5
Rail	41	3	5	3
Short sea shipping	23	3	10	1

Table 9.8 Subcriteria for the modal choice variable "security".

Transport mode	Theft	Sabotage	Terrorism
Road	2	2	3
Rail	3	3	4
Short sea shipping	4	4	4

In the PROMETHEE method, next to the evaluation table, additional information concerning the importance of the various criteria (weights) and the importance of a difference in the evaluations (represented by a preference function) needs to be given. The weights were chosen as obtained by the survey described before. The subcriteria of security were given an equal weight. A V-type preference function was selected for the criteria transport cost and transport duration (this means that the preference increases as the difference in costs increases), while for the ordinal criteria (security, flexibility and reliability) the normal preference function was chosen (in the case of a difference in values, there will be a preference for one alternative compared to the other).

Based on these data, road transport is preferred to the other modes (see Figure 9.8). Despite the price difference, the other modal choice variables seem to have a big impact. This corresponds to situations in practice; the shipper chooses road transport despite the cost difference. However, rail also has a high score.

This is also clearly shown in the GAIA-plane (see Figure 9.9). In this graph, the different criteria are represented by axes and the alternatives are projected into a two-dimensional plane. The alternatives “Road” and “Rail” are generally seen as the best one, since the dark gray axis (the decision stick) points towards the most preferred alternative. Short sea shipping scores the best for the criterion “transport cost”.

If security is not taken into account, the results shown in Figure 9.10 are obtained.

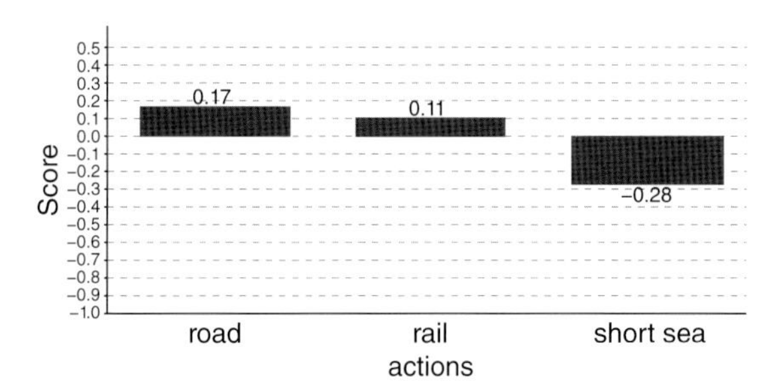

Figure 9.8 PROMETHEE II ranking.

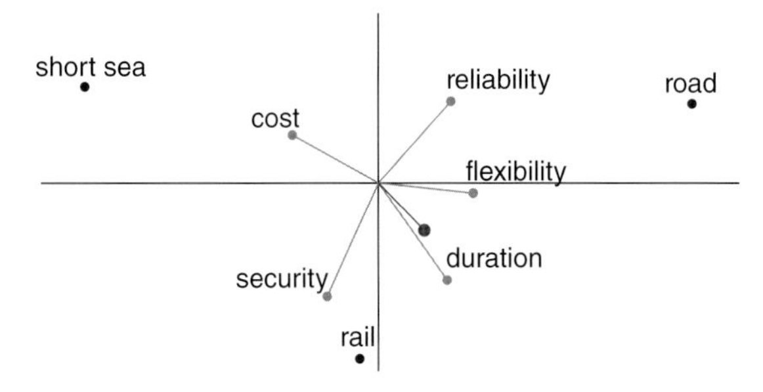

Figure 9.9 GAIA plane.

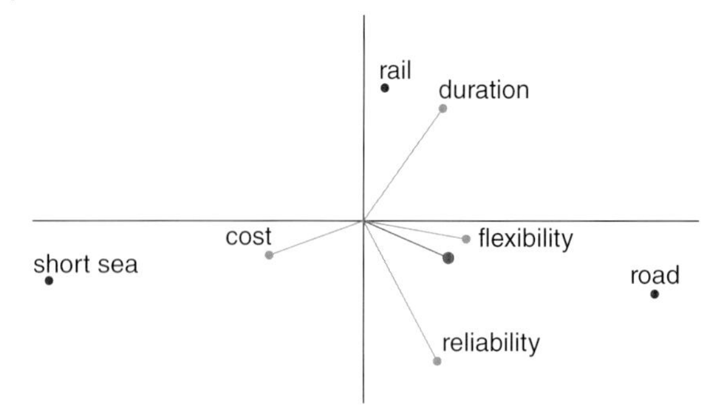

Figure 9.10 GAIA plane (without security).

This example shows that security has an important influence on the transport decision. Without the inclusion of security aspects, rail transport would not have been considered as a sensible choice as other modes are more competitive for the majority of criteria.

9.6 Conclusion

The first part of this chapter consisted of a literature review in which the different hazmat transport criteria and some existing risk analysis systems were being discussed. The lessons learned from the literature review eventually led to the refinement of the existing methods and the development of an adapted approach that used historical accident data to calculate the risks of multimodal hazmat transport. In this revised approach one of the general principles is that the risk of a catastrophic hazmat incident can be divided into two parts, which can both be validated on the basis of accident data: (i) the calculation of the general probability of the occurrence of an accident based on international accident data of transport of hazardous materials – this is the basis for the global risk map and (ii) the calculation of the local probability of the occurrence of an accident based on accident data and infrastructure parameters of the complete available freight transport in Flanders – this is the basis for the local risk map. The ratio between these two results in a locality parameter, which represents the local specific circumstances that can lead to an accident. This new evaluation framework makes it possible to estimate the risks of hazmat transport along a specific route for transport by road, rail, inland navigation and even pipelines. The results allow comparison of those different transport modes based on the average frequency of release and the different risk parameters, as well as the calculation of the number of potential victims along the route. The global risk maps give a good indication of the overall risk due to the transport of dangerous goods, while the underlying risk maps for each of the four types of hazardous substances display the individual contributions of each type to the overall risk. Finally, the methodology also allows estimation of the consequences of a possible increase or decrease of the transported volume on a route, or the effect of an increase or decrease in the number of people exposed along the route. It can be concluded that, especially for unintended incidents such as traffic accidents, it is better to use a QRA approach to determine risk-sensitive segments for hazmat transport along the transport network. However, due to the scarcity of historical intented incident data, QRA may not be applicable. A MCA method to include security in the modal choice is proposed.

In the case study, the transport decisions of a shipper were examined for a specific route on the basis of multicriteria analysis. Various criteria, such as transport costs, transport duration, reliability, flexibility and security, were taken into account in this multicriteria analysis. The evaluation of three different alternatives (road, rail and short sea shipping) resulted in a ranking of these alternatives. In order to calculate the transport costs, the "total logistic cost" method was used. Based on these total logistic costs, road transport is the most expensive transport, followed by intermodal transport by rail. Intermodal transport using short sea shipping proved to be the least expensive option. The analysis showed that taking into account security aspects considerably changed the analysis.

Acknowledgments

The development of this methodology was realized with the financial support of the Flemish Government, Department Environment, Nature and Energy–Safety Reporting (LNE–dienst VR). We also want to thank our consortium partners A. Debeil (Safety Advisors), E. Vreys (Möbius) and S. De Schrijver (Möbius) for their cooperation during this project.

Bibliography

Abkowitz, M. and Cheng, P. (1988) Developing a risk cost framework for routing truck movements of hazardous materials. *Accident Analysis and Prevention*, **20** (1), 39–51.

Abkowitz, M., Lepofsky, M., and Cheng, P. (1991) Selecting criteria for designating hazardous materials highway routes. *Transportation Research Record*, **1333**, 30–35. TRB NRC, Washington DC, USA.

Behzadian, M., Kazemzadeh, R.B., Albadvi, A., and Aghdasi, M. (2010) PROMETHEE: a comprehensive literature review on methodologies and applications. *European Journal of Operational Research*, **200** (1): 198–215.

Belton, V. and Stewart, T.J. (2002) *Multiple Criteria Decision Making. An Integrated Approach*, Kluwer Academic Publishers, Boston/ Dordrecht/ London.

Beynon, M.J. and Wells, P. (2008) The lean improvement of the chemical emissions of motor vehicles based on their preference ranking: a PROMETHEE uncertainty analysis, OMEGA. *International Journal of Management Science*, **36** (3), 384–394.

Blauwens, G., Janssens, S., Vernimmen, B., and Witlox, F. (2002) "The importance of frequency for combined transport of containers", Research Paper 2002–030, UFSIA-RUCA Faculty TEW, University of Antwerp, 21 blz.

Brans, J.P. (1982) L'ingéniérie de la décision. Elaboration d'instruments d'aide à la décision. Méthode PROMETHEE, Université Laval, Québec, Canada.

Brans, J.P., Mareschal, B. (1994) PROMCALC & GAIA: a new decision support system for multicriteria decision aid. *Decision Support Systems*, **12**, 297–310.

Brans, J.P., Mareschal, B., and Vincke, P. (1986) How to select and how to rank projects: the PROMETHEE method for MCDM. *EJOR*, **24**, 228–238.

Brans, J.P. and Vincke, P. (1985) A preference ranking organisation method: the PROMETHEE method for MCDM. *Management Science*, **31** (6), 647–656.

Clark, R.M. and Besterfield-Sacre, M.E. (2009) A new approach to hazardous materials transportation risk analysis: decision modeling to identify critical variables. *Risk Analysis*, **29** (3), 344–354.

Dagdeviren, M. (2008) Decision making in equipment selection: an integrated approach with AHP and PROMETHEE. *Journal of Intelligent Manufacturing*, **19**, 397–406.

De Brucker, K., Verbeke, A., and Macharis, C. (2004) The applicability of multicriteria-analysis to the evaluation of intelligent transport systems (ITS), in *Economic Impacts of Intelligent Transportation Systems: Innovations and Case Studies. Research in Transport Economics*, (eds E. Bekiaris and Y.J. Nakanishi), Elsevier Ltd., Amsterdam, 8: 151–179.

GAO (2009) Freight Rail Security: Actions Have Been Taken to Enhance Security, but the Federal Strategy Can Be Strengthened and Security Efforts Better Monitored. GAO-09-243 April 21, 2009.

Garson, G.D. (2010) Multiple Regression, from Statnotes: Topics in Multivariate Analysis, http://faculty.chass.ncsu.edu/garson/pa765/statnote.htm (accessed 27 September 2010).

Hesketh, D. (2010) Weaknesses in the supply chain: who packed the box? *World Customs Journal*, **4** (2), 3–20.

Hintsa, J. (2011) Post-2001 supply chain security–impacts on the private sector. Doctoral thesis. HEC University of Lausanne. 28.1.2011.

Huang, B., Cheu, R.L., and Liew, Y.S. (2004) GIS and genetic algorithms for HAZMAT route planning with security considerations. *International Journal of Geographical Information Science*, **18** (8), 769–787.

IRU (2005) Voluntary Security Guidelines for Managers, Drivers, Shippers, Operators Carrying Dangerous Goods and Customs–Related Guidelines. Geneva 2005.

JTP (2010) Research Methods II: Multivariate Analysis. *Journal of Tropical Pediatrics*, http://www.oxfordjournals.org/our_journals/tropej/online/ma.html (accessed 27 September 2010).

Kumar, A., Sokhansanj, S., and Flynn, P.C. (2006) Development of a multicirtieria assessment model for ranking biomass feedstock collection and transportation systems. *Applied Biochemistry and Biotechnology*, **129** (1–3), 71–87.

Leonelli, P., Bonvicini, S., and Spadoni, G. (2000) Hazardous materials transportation: a risk-analysis-based routing methodology. *Journal of Hazardous Materials*, **71**, 283–300.

Lepofsky, M. and Abkowitz, M. (1993) Transportation hazard analysis in integrated GIS environment. *Journal of Transport Engineering*, **119**, 239–254.

LNE (2009) TWOL-project: Berekeningsmethodieken en Veiligheidsmaatregelen voor Transport Gevaarlijke Stoffen: Eindrapport. Vlaamse Overheid: Departement Leefmilieu, Natuur en Energie (Dienst Veiligheidsrapportering).

LNE (2010) Risicoanalysesysteem voor het Transport van Gevaarlijke Stoffen: Eindrapport. Vlaamse Overheid: Departement Leefmilieu, Natuur en Energie (Dienst Veiligheidsrapportering).

Lord, D. and Mannering, F. (2010) The statistical analysis of crash-frequency data: a review and assessment of methodological alternatives. *Transportation Research Part A*, **44**, 291–305.

Lord, D., Washington, S.P., and Ivan, J.N. (2004) Statistical challenges with modeling motor vehicle crashes: understanding the implications of alternative approaches. Center for Transportation Safety. Texas Transportation Institute.

McCullagh, P. and Nelder, J.A. (1989) *Generalized Linear Models*, 2nd edn, Chapman and Hall, London.

Macharis, C., Brans, J.P., and Mareschal, B. (1998) The GDSS PROMETHEE procedure: a PROMETHEE-GAIA based procedure for group decision support. *Journal of Decision Systems*, **7**, special issue, 238–307.

Macharis, C. and Geudens, T. (2009) Reviewing the use of multi-criteria decision for the evaluation of transport projects: time for a multi-actor approach, 34 pp.

Macharis, C., Heugens, A., and van Lier, T. (2008) "Het belang van perceptie in de modale keuze: een case study", Vervoerslogistieke werkdagen, Deurne (Nederland).

Macharis, C., Verbeke, A., and De Brucker, K. (2004) The strategic evaluation of new technologies through multicriteria analysis: the advisors case, in *Economic Impacts of Intelligent Transportation Systems: Innovations and Case Studies. Research in Transport Economics*, (eds E. Bekiaris and Y.J. Nakanishi), Elsevier Ltd., Amsterdam, 8.

Mohamadabadi, H.S., Tichkowsky, G., and Kumar, A. (2009) Development of a multi-criteria assessment model for ranking of renewable and non-renewable transportation fuel vehicles. *Energy* **34**, 112–125.

Norojono, O. and Young, W. (2003) A stated preference freight mode choice model. *Transportation Planning and Technology*, **26**, 195–212.

Panwhar, S.T., Pitt, R., and Anderson, M.D. (2000) Development of a GIS-based Hazardous Materials Transportation Management System: A Demonstration Project. UTCA Report 99244. Tuscaloosa, Alabama: University Transportation Center for Alabama.

Reniers, G., Gorrens, B., De Jongh, K., Van Leest, M., Lauwers, D., and Witlox, F. (2009) De ontwikkeling van een methodiek voor de risicobepaling van transporten van gevaarlijke stoffen: Een eerste toepassing voor Vlaanderen. *Bijdragen Vervoerslogistieke Werkdagen*, **2009**, 259–272.

Saccomanno, F.F. and Chan, Y.W. (1985) Economic evaluation of routing strategies for hazardous road shipments. *Transportation Research Record*, **1020**, 12–18. TRB NRC, Washington DC, USA.

Sheffi, Y. (2001) Supply chain management under the threat of international terrorism. *International Journal of Logistics Management*, **12** (2), 1–11.

Shinghal, N. and Fowkes, T. (2002) Freight mode choice and adaptive stated preferences. *Transportation Research Part E: Logistics and Transportation Review*, **38** (5), 367–378.

Tongzon, J. (1995) Determinants of port performance and efficiency. *Transportation Research*, **29A** (3), 245–252.

Tsamboulas, D. and Moraiti, P. (2008) Identification of potential target locations and attractiveness assessment due to terrorism in the freight transport. *Journal of Transportation Security*, **1** (3), 189–207.

Tsamboulas, D., Yiotis, G., and Panou, K. (1999) Use of multicriteria methods for assessment of transport projects. *Journal of Transportation Engineering*, **25** (5), 404–414.

Tuzkaya, U.R. (2009) Evaluating the environmental effects of transportation modes using an integrated methodology and an application. *International Journal of Environmental Science and Technology*, **6** (2), 277–290.

United Nations Economic and Social Council (2008) Economic Commission for Europe Inland Transport Committee, Working Party on the Transport of Dangerous Goods. ECE/TRAN/WP.15/197.

Van Geirt, F. and Nuyts, E. (2006) Handleiding bij het gebruik van regressiemodellen voor ongevalrisico's. RA-2006-89. Steunpunt Mobiliteit en Openbare Werken – Spoor Verkeersveiligheid.

Vannieuwenhuyse, B. and Germis, J. (2002) "Promodi resultatenrapport", VEV-CIB-enquête en onderzoek a.d.h.v. interactieve webapplicatie, eindrapport, Antwerp-Leuven, October 2002.

Vannieuwenhuyse, B., Gelders, L., and Pintelon, L. (2003) An online decision support system for transportation mode choice. *Logistics Information Management*, **16** (2), 125–133.

Zhu, S.-H. and Anderson, N.H. (1991) Self-estimation of weight parameter in multi attribute analysis. *Organizational Behavior and Human Decision Processes*, **48**, 36–54.

Zuur, A.F., Ieno, E.N., Walker, N.J., Saveliev, A.A., and Smith, G.M. (2009) *Mixed Effects Models and Extensions in Ecology with R*, Springer, New York, NY, 574 pp.

10
Metaheuristics for the Multimodal Optimization of Hazmat Transports

Kenneth Sörensen, Pablo Maya Duque, Christine Vanovermeire, and Marco Castro[1)]

10.1
Introduction

An increase of multimodal transportation has been hailed as one of the most important steps towards a more sustainable global economy. By coining the term *comodality* (European Commission, 2006) the European Commission has indicated that it supports a view on transport in which the different modes are not competing, but rather complementary. However, multimodal transportation, especially that of hazardous materials, poses a large number of challenges. From a technical point of view, the appropriate safety and security measures to take depend on the type of material that is being transported, the route it should follow from its origin to its destination and the transport mode that is being used.

Besides these technical questions, the *organizational* challenges involved in *planning* hazmat transports are also particularly pertinent. The field of OR/MS (operations research/management science) disposes of a large arsenal of tools and techniques for transportation planning, and is therefore the research field most able to assist in solving such organizational problems. Transport-planning problems are generally stated as *optimization problems*, in which the values of a set of *decision variables* (e.g., which transport modes to use for a certain shipment and in which order to visit a set of customers) is determined in such a way that some *objective function* (e.g., total cost or total risk) is minimized (or maximized), subject to some *constraints* (e.g., material should be delivered on time). For many such transportation problems, efficient optimization models and methods have been developed. However, the additional challenges posed by the transportation of hazardous materials should not be underestimated. As became clear from the previous chapters, hazardous materials are often transported in a multimodal way, i.e. they encounter several transshipment points on their way from source to destination (see Figure 10.1). Transshipment requires handling of the cargo, which yields an additional risk. Moreover, planning of multimodal transport is

1) kenneth.sorensen@ua.ac.be, pablo.mayaduque@ua.ac.be, christine.vanovermeire@ua.ac.be, marco.castro@ua.ac.be.

Security Aspects of Uni- and Multimodal Hazmat Transportation Systems, First Edition. Edited by Genserik L.L. Reniers, Luca Zamparini.

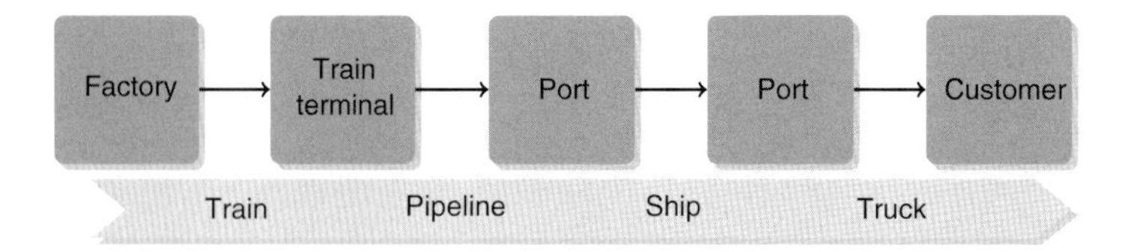

Figure 10.1 A multimodal supply chain: decisions need to be made on which transshipment points to use, which transport modes to use, and how exactly to send a shipment from the factory to the customer.

considerably more difficult than unimodal transport planning, because additional decision variables make the problem larger and more complex.

A complete overview of optimization problems that can arise in the multimodal transportation of hazardous materials is beyond the scope of this chapter. However, optimization problems typically divide into two categories:

- *Operational problems* involve short-term decisions, such as the exact way in which a shipment X should be transported from factory A to customer B, i.e. using which transport mode(s), which transshipment point(s), which route(s), etc.

- *Tactical/strategic problems* are concerned with long-term decisions. These include the network layout of the intermodal infrastructure (where should terminals be built, pipelines be laid, canals be developed, etc.).

It is interesting to note that optimization problems often combine aspects from both categories, and that information from the operational level is necessary to determine the best strategic decisions. For example, the decision which ports to ship through, is of course dependent on the way in which the goods are shipped through these ports. Optimization problems in which different decisions on different levels are simultaneously taken, are called *multilevel* problems.

Optimization *methods* can also be divided into two categories: exact and heuristic. The difference is that *exact* methods come with a guarantee to find the *optimal* solution, i.e. the solution with the lowest[2] objective function value (e.g., cost) from the set of all feasible[3] solutions. The main problem that exact methods face is that most interesting optimization problems, especially those in multimodal transportation are computationally *difficult*. This means that their computing time rises exponentially with the size of the problem. Practically, this most often results in prohibitively large computing times that can range from several hours, to several days, years or even centuries for all but the smallest of problems. In other words, when applying an exact method to a realistic multimodal transportation planning problem, there is no guarantee that the method will produce a solution in a reasonable amount of time, or even within the lifetime of the transportation planner.

2) For a minimization problem.
3) A solution is called feasible if it satisfies all the constraints set forward in the optimization problem.

Heuristic methods, often called *heuristics*, are defined by the fact that they are not exact methods, i.e. they do not come with a guarantee that the optimal solution will be found. Instead they attempt to find a solution that is as good as possible, in a computing time that is as small as possible. Heuristics are more flexible than exact methods, in that the amount of time allocated to the method may be determined by the application: some applications require almost-instant decisions (e.g., rerouting a car when it has taken a wrong turn is done by the satnav device in a few seconds at most), whereas others may be allowed hours or even days (e.g., strategic decisions, such as determining the location of intermodal terminals).

Before the 1980s, heuristics were usually very simple procedures or rules-of-thumb, based on intuition and common sense. As a result, heuristics were often regarded as inferior to exact methods, which was recognized in the following quote made by Fred Glover (1977).

> "[Exact] algorithms are conceived in analytic purity in the high citadels of academic research, heuristics are midwifed by expediency in the dark corners of the practitioner's lair, . . . and are accorded lower status."

The advent of more advanced heuristic methods, and the development of the field of *metaheuristics* has completely changed this picture.

10.2 Metaheuristics

A *metaheuristic* is defined as a "high-level problem-independent algorithmic framework that provides a set of guidelines or strategies to develop heuristic optimization algorithms" (Sörensen and Glover, to appear). The same term is also used to refer to a problem-specific implementation of a heuristic optimization *algorithm* according to the guidelines expressed in such a framework. The term combines the greek prefix "meta" (beyond, in the sense of high-level) with "heuriskein" (to search) and was coined in (Glover, 1986).

Figure 10.2 shows the positioning of metaheuristics, simple heuristics and exact methods with respect to computing time and solution quality. The aim of metaheuristics is to obtain a solution quality that is as close as possible to that of the optimal solution, in a computing time that is much lower than that of exact methods.

The defining characteristic of a metaheuristic (framework) is that the resulting methods are always heuristic in nature, i.e. metaheuristics are never exact methods that will find the optimal solution in a finite number of steps.[4] Metaheuristics frameworks find their inspiration in many different fields, some seemingly unre-

4) Recently, the fields of metaheuristics and exact methods are drawing closer together, and many methods have been proposed that integrate exact methods into metaheuristics. Nonetheless, a metaheuristic algorithm is always a heuristic.

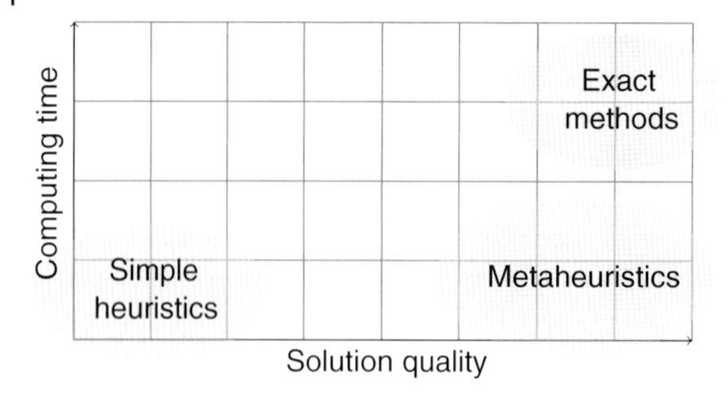

Figure 10.2 Simple heuristics, exact methods and metaheuristics.

lated to optimization. In this vein, metaheuristics have been proposed that mimick natural evolution (evolutionary algorithms), or the behavior of a group of foraging ants (ant-colony optimization). One of the first metaheuristics to be proposed, *simulated annealing* (Kirkpatrick, Gelatt, and Vecchi, 1983), models the optimization process after the cooling of a crystalline solid. Other metaheuristic frameworks stick closer to the traditional views of optimization (e.g., tabu search, variable neighborhood search, . . .).

Although many (possibly hundreds) of "different" metaheuristics methods have been proposed, they can be divided into three categories, based on the fundamental engine that drives the search for good solutions.

- *Local-search* metaheuristics work by *iteratively making small changes* to a *single solution*, called the *current* solution. Some notable examples of metaheuristics in this category include the following. *Iterated local search* (Lourenco, Martin, and Stützl, 2003) uses large random jumps to explore different solutions. *Tabu search* (Glover, 1989, 1990; Glover and Laguna, 1993) uses memory structures to remember what the search has done in the past and exploit this to improve the search in the future. *Variable neighborhood search* uses different types of "small changes" (called neighborhoods) at the same time.
- *Constructive* metaheuristics operate by iteratively *building solutions from their constituting parts*, thereby operating on a single, *incomplete solution*. Examples of metaheuristics in this category are *greedy randomized adaptive search procedure* (GRASP [Feo and Resende, 1995], that adds some randomness to a greedy algorithm to allow it to find more and better solutions) and *ant-colony optimization* (Dorigo, Maniezzo, and Colorni, 1996; Dorigo, Birattari, and Stützl, 2006, 2006), that models the optimization process after the food-searching behavior of a colony of ants. The latter has received widespread attention in the popular press (see, e.g., The Economist, 2010), probably as a result of the attractiveness of the metaphor and has also spawned a wide range of metaheuristics based on the behavior of other insects (fleas, termites, bees, . . .).
- *Population-based* metaheuristics find better solutions by *maintaining a set of solutions* (often called the *population*) and iteratively *combining solutions into new*

solutions. Most metaheuristics in this class belong to the category of *evolutionary algorithms*, that mimic the process of natural evolution first described by Charles Darwin. Starting with the work of Holland (1975), a large number of related metaheuristics have been proposed since the early 1980s, including *genetic algorithms* (Goldberg *et al.*, 1989), evolution strategies (Beyer and Schwefel, 2002), and genetic programming (Koza, 1992).

It should be noted that these classes of metaheuristics are not mutually exclusive, and that many (if not most) metaheuristic algorithms combine elements from different classes. *Memetic algorithms* (Moscato, 1989), for example, combine local search improvement heuristics with evolutionary crossover operators. In the metaheuristics literature, there is a growing tendency to view these different classes as components that can be arbitrarily combined in novel ways. Moreover, problem knowledge is important in determining which method will work well on which problem.

Regardless of the underlying optimization strategy (local search, construction or evolution), metaheuristics have been hugely successful, and have become the method of choice for solving optimization problems that are large, complex, or both. Where exact methods fail to deliver the tools necessary to solve real-life optimization problems, metaheuristics have filled the gap and are now the default choice for all but the simplest of problems. Indeed, most commercial packages for solving scheduling or routing problems include some metaheuristic components (Sörensen, Sevaux, and Schittekat, 2008). Exact methods, although useful from a theoretical point of view, lack the flexibility to be used in real-life applications and are only used as one of the components of an otherwise heuristic algorithm.

A full description of all metaheuristics is well beyond the scope of this chapter. The interested reader is referred to Michalewicz and Fogel (2004), and Talbi (2009).

10.3 Characteristics of Multimodal Hazmat Transportation Optimization Problems and the Case for Using Metaheuristics

It is not difficult to see that multimodal hazmat transportation optimization problems generally classify as both large and complex. For this reason, metaheuristics present the most appropriate approach to tackle such problems.

In today's global economy, most companies that produce and transport hazardous materials are large and operate on a worldwide basis. As a result, the supply chains of such companies span the globe, and – in order to remain competitive – the distribution of goods from production facilities to customers needs to operate like a clockwork in which all logistics operations are closely aligned. Additionally, many different decisions need to be taken, both short-term operational ones and long-term strategic and tactical ones. Real-life instances of multimodal hazmat transportation problems generally consist of a lot of transportation orders. Incidents may occur anywhere in the supply chain and generally have a complicated impact.

Moreover, the amount of time that is available to solve problems of hazmat transportation is not unlimited. This is especially true for operational models, in which short term decisions are taken, e.g., on the exact truck in which a shipment is to be transported from the warehouse to the customer, or the order in which to visit customers in the milk run of a truck on a given day.

Perhaps surprisingly, there is also a preference for algorithms for strategic decisions to be as fast as possible. The reason is that such decisions are typically not taken in a one-shot way, but by using an integrated decision tool in which the user can manually test certain assumptions (e.g., would opening a new warehouse in this location be beneficial? What would be the effect of building a new pipeline between points A and B?). For a given user-input scenario, the optimization algorithm integrated into the decision support tool should be able to quickly determine a "good enough" answer.

In some situations, models even need to be solvable in real time. Zhang, Hodgson, and Erkut (2000) discuss the case of evacuation guidelines, in which a speedy and accurate routing that reacts to the substance transported and the weather conditions (e.g., airborne toxics are more dangerous in the case of windy weather) can save many lives.

The sheer size and complexity of most realistic multimodal hazmat transportation optimization problems, as well as the diversity of problems encountered in real life, call for the use of metaheuristic optimization tools. There are, however, two important additional characteristics that further strengthen the need for heuristic optimization strategies: the fact that *different types of decisions* need to be taken on different levels simultaneously, as well as the fact that hazmat transportation optimization problems have *multiple objectives*. These are discussed in the following sections.

10.3.1 Multilevelness

Multimodal hazmat transportation has some advantages over its unimodal counterpart. Not only is it found to be increasingly important (Erkut, Tjandra, and Verter, 2007), using different transport modes offers interesting possibilities to reduce risk while still remaining competitive. Indeed, as each mode of transport has different cost and service profiles, as well as varying risk factors (Rondinelli and Berry, 2000; Verma and Verter, 2007), combining those different modes offers additional opportunities for optimization.

However, allowing hazmat products to travel on different transport modes also renders the underlying optimization problem more difficult. The optimal routes (how much of which substance should be transported via which way and using which mode) cannot be determined if the risk of each route is not known. The risk cannot be calculated if the mode(s) used and the amount of the substance transported is not yet determined. Such intertwining of problems demands a multilevel modeling approach.

Multilevel modeling implies that the problem at hand is divided into a master problem and linked subproblem(s) and that decisions taken at each level influence the value (e.g., the cost) of decisions taken at the other level(s). In the case we described above, we try to find a route at minimum cost and risk (master problem), for which the subproblem is the determination of the risk for each route. The latter problem can be computationally difficult by itself. Carotenuto, Giordani, and Ricciardelli (2007) and Zhang, Hodgson, and Erkut (2000) for example, use map algebra techniques from Geographic Information Systems to determine the risk for a given route on the basis of the transport mode, as well as the substance transported, the environmental conditions and the population exposure.

Other multilevel problems abound in multimodal hazmat transportation problems. For example, the optimal location of intermodal terminals depends heavily on the goods that will be transported through the network. The cost of a solution to a terminal location problem (i.e., a choice of terminals to locate) therefore can only be determined by solving another optimization problem: how will the goods be routed through the network given a set of open terminals (Sörensen, Vanovermeire, and Busschaert, 2011).

Walshaw (2004) gives an overview on metaheuristics that work well on multilevel problems and reports that a wide variety of metaheuristics have been successfully applied to such problems. He finds that partitioning into subproblems and thus creating approximations, is easier to implement and increases speed. Generally multilevel problems are either solved sequentially or simultaneously (Nagy and Salhi, 2007). In the sequential approach, the master problem is solved first, followed by the subproblem. In the simultaneous approach, both problems are solved at the same time. It is generally acknowledged that the simultaneous approach should be preferred.

10.3.2 Multiobjectivity

Rondinelli and Berry (2000) state that failing to sufficiently manage risk leads to a negative environmental impact, but can also result in a loss of competitive advantage. Indeed, environmental hazards also contain a business risk, and can have a clear and direct financial impact (e.g., fines issued by the government, clean-ups, . . .) or effects that are more difficult to measure (e.g., damage to corporate image). Rondinelli and Berry (2000) argue that rather than complying with constraints imposed by the government, companies need to proactively minimize the risk of environmental damage in order to reduce their business risk. If we translate this argument to the field of transportation and logistics, it implies that companies should design their logistics systems in such a way that the transportation risk is minimized. Similarly, operational decisions should be taken in the same way: if a company has the choice between transporting hazardous materials through an urban center or not, it should take the latter option.

However, two aspects that complicate matters considerably can arise. First, the risk involved in taking certain decisions needs to be quantified. This is difficult and has not been done in a uniform way. Erkut and Verter (1998) report five different ways to quantify risk in the literature, none of them presenting a significantly better choice than the other, since the "best" risk quantification method is dependent on the strategies and preferences of the stakeholders. Importantly, redefining risk as a cost is not mentioned as a possibility of risk quantification.

It is indeed often undesirable (or even impossible) to express security aspects in terms of cost. This renders the optimization problem of multimodal hazmat transports a *multiobjective* one, in which several conflicting objectives (total cost and total risk are obvious, but other objectives are possible, such as environmental impact, etc.) have to be balanced. In such problems, there is not a single solution that is "better" than all others: the solution (transport plan) with lowest total risk will generally not be the same as the one with lowest cost.

As a result, the notion of "better" needs to be abandoned, as a solution can be better with respect to one objective than another solution, but worse with respect to another. In the field of *multiobjective optimization*, this notion is usually replaced by the (weaker) concept of *domination*. A solution is said to dominate another solution if it is better with respect to all objectives.[5] The solutions that are not dominated by any other solution are said to be in the *Pareto set* (or Pareto front). This is shown in Figure 10.3, in which we search for solutions that simultaneously minimize objectives (e.g., total cost and total risk). All solutions on the Pareto set (represented as dots) are mutually nondominating, i.e. it is impossible to say that one is better than the other, unless some extra information on the preferences of the decision maker is added. One decision maker may be risk averse and prefer solutions that have low risk, whereas another may want to find a solution with low cost, regardless of the risk involved. However, whatever the risk preferences of the different stakeholders may be, a solution that is dominated will never be chosen over one that is in the Pareto set.

Given the complexity of multiobjective optimization problems, it is generally impossible to find all solutions on the Pareto set. The common answer to this

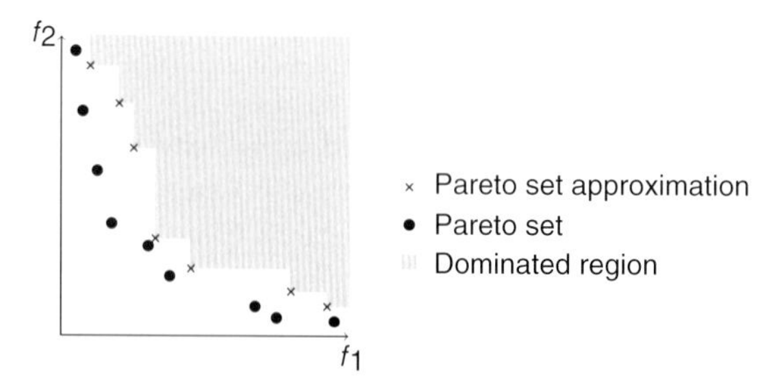

Figure 10.3 Multiobjective optimization (two objectives).

5) More correctly, solution a dominates solution b if it is a at least as good as b on all objectives and better on at least one.

problem is to use a heuristic procedure to *approximate* the Pareto set as well as possible. The set of solutions returned by this heuristic procedure should be mutually nondominating and is called the *Pareto set approximation*. This is also shown in Figure 10.3, in which solutions in the Pareto set approximation are represented by a (symbol. Solutions that are dominated by one or more solutions in the Pareto set approximation are not considered interesting, because there exist solutions that are better with respect to every objective. Such solutions are said to be in the *dominated region*, indicated as the shaded region in Figure 10.3 and may be safely discarded by the algorithm.

Approaching the optimal Pareto set is the core of most recent multiobjective approaches. One possible approach consists in aggregating the objectives into a scalar function (a weighted average of the objective function values) and solving the resulting single-objective optimization problems. Although this approach is frowned upon for theoretical reasons,[6] it is still often used.

Marler and Arora (2004) categorize approaches for multiobjective optimization depending on how the preferences of the decision maker (i.e., the relative importance of the objectives) are handled. The methods are divided into three major categories: methods with *a priori* articulation of preferences in which the user indicates the relative importance of the objective before running the optimization algorithm; methods with *a posteriori* articulation of preferences in which the algorithm generates a set of equivalent solutions from which one is selected by the decision maker; and methods with no articulation of preferences. According to Jones, Mirrazavi, and Tamiz (2002), most of those standard methods for multiobjective problems are based on basic linear and integer programming, leaving aside complicating factors that are common when dealing with real problems. Thus, those models do not consider issues such as nonlinearity, stochasticity, nonstandard constraints and feasibility conditions.

However, the increase in computing power and the development of more appropriate algorithms have made better modeling of such conflicting situations possible. Regarding the progress in designing better algorithms, the use of metaheuristics has played an important role. As pointed out by Jones, Mirrazavi, and Tamiz (2002), the use of metaheuristic approaches has been supported not only by the increase in computing capacity, but also by the transferral effect of advances on the application of these algorithms to single-objective problems, as well as the growing awareness of the importance of considering multiple objectives in diverse disciplines. Indeed, multiobjective metaheuristics have brought multiple objective programming into disciplines such as engineering and medicine, for which the use of these models was constrained by the limitations of standard approaches.

Evolutionary algorithms have become the predominant method for exploring the optimal Pareto optimal set in multiobjective optimization. This is due mainly

6) Most importantly, because it can only find *supported* solutions, i.e. solutions that are the best with respect to at least one setting of the scalarizing weights, whereas most optimization problems also have Pareto-optimal solutions that are not supported.

to the fact that they are population-based, which allows them to find multiple solutions in a single run of the algorithm. Therefore, some of the most well-known algorithms for solving multiple-objective problems belong to this class, such as NSGA-II, PAES and SPEA2 proposed by Deb *et al.* (2002), Knowles and Corne (1999), and Zitzler *et al.* (2001), respectively.

An interesting approach is the one by Verma (2009), who created a model that proposes a number of solutions that are nondominating with respect to both cost and people exposure. Pradhananga, Taniguchi, and Yamada (2010) notes that another advantage of generating a Pareto front is that it offers several solutions that are equally "good" to stakeholders that have different objectives. The authors therefore develop a biobjective metaheuristic based on ant-colony optimization.

10.4 Metaheuristics for Multimodal Hazmat Transportation

In the previous section, we have made a strong case for the application of metaheuristics to the optimization of multimodal hazmat transport optimization problems. The flexibility of metaheuristics, their speed and the fact that they have been shown to be able to handle multilevel and multiobjective optimization problems, strongly motivates the development of these techniques to solve the interesting real-life problems that appear in the multimodal transportation of hazardous materials.

Notwithstanding this fact, the literature on metaheuristics for multimodal hazmat transport optimization is remarkably sparse. In fact, we were unable to locate a single contribution in this field. However, some interesting papers have appeared in recent years on (i) metaheuristics for multimodal transportation, and (ii) metaheuristics for hazmat transportation. The intersection of these research subdomains present a tremendous challenge for future research.

10.4.1 Metaheuristics for Multimodal Transportation

As mentioned before, multilevel modeling usually decomposes the problem at hand into a master problem with linked subproblems. This approach is followed by Castelli, Pesenti, and Ukovich (2004) when proposing a constructive heuristic method for scheduling multimodal transportation networks. In particular, this work focuses on peak periods of demand during the day where the main objective is to minimize operational costs, serving–at the same time–as many customers as possible. The authors deal with the *transfer coordination problem* (see Knoppers and Muller, 1995) with some assumptions on the means of transport. They define a mathematical model for timetabling a transportation system, and they also present a (Lagrangian) relaxed version of the problem decomposing it into a set of single-line subproblems. The constructive procedure then schedules a single line at a time. In view of the fact that this approach turns out to be intractable for real systems, the authors propose a Lagrangian-based algorithm that iteratively estimates the optimal values of the decision variables and of the Lagrangian multipliers based on an economical interpretation of their real meaning. In this way, the algorithm heuristically solves the problem of scheduling a single line. Scheduling the entire system is carried out by scheduling one line after another. The order in which the lines are scheduled should be based on the operator's own experience, which implies that this procedure is interactive.

Jansen *et al.* (2004) describe a successful implementation of an operational planning system for a transportation company based in Germany. The objective of the planning process is to provide a cost-efficient transportation plan for a given set of orders, taking a large number of constraints into account, and considering two modes of transportation: rail and road. Due to the complexity and large scale of the problem, it is decomposed into a number of subproblems and an efficient and effective algorithm is developed for each subproblem. This decomposition allows for a flexible implementation, in which initial solutions are easy to construct and partial solutions can be retrieved when algorithm is stopped. The system involves the complementary use of exact and heuristic approaches. Thus, path-search and standard network-flow algorithms, and local-search and global-search heuristics are used at different levels of the solution procedure. Although the savings obtained by using the system are difficult to measure, the total cost savings estimated are at least at 5%, which leads to a decrease in cost of the order of several million dollars per year.

In Galvez-Fernandez *et al.* (2009) a graph theory-based approach for a time-dependent multimodal transport problem (TMTP) is proposed. The authors present a graph structure called a *transfer graph* as an abstraction of a multimodal transport network. The transfer graph separates all transport modes in unimodal networks by defining a unimodal transport network for each transport mode. From this model, a simplified structure arises, called the *relevant graph*, from which the shortest path between origin and destination can be computed more easily. First, part of the calculations are performed and stored in advance. These precalculations consist of computing the shortest path between transfer vertices at any possible departure time. Then, a variation of Dijkstra's algorithm is proposed, with some modifications to support time dependence. Based on the information previously stored, the relevant graph is built. Finally, the shortest path between a source and a destination node is computed from this new abstraction. This approach is experimentally validated obtaining better results in terms of CPU time than other related approaches. However, the work of Galvez-Fernandez *et al.* (2009) suffers from limited scalability due to memory-space requirements. Ayed, Habbas, and Khadraoui (2009) describe the use of an ant-colony system to explore the graph in order to find shortest paths between transfer vertices, thus reducing considerably the memory-space complexity compared with Dijkstra's algorithm. Nevertheless, computing time is increased considerably by this approach. Therefore, Ayed *et al.* (2010) present a hybrid approach that is a more balanced solution in terms of computation time and memory-space requirements. This approach creates an intermediary structure called an *abstract graph* instead of computing the precalculations mentioned before. Then, a *relevant intergraph* is built from which the shortest path is calculated. The authors implemented an ant-colony system to create the abstract graph.

Yamada *et al.* (2009) deal with the problem of investment planning in developing multimodal freight transport networks. In this problem a set of possible actions is considered that can include actions such as improving existing infrastructure or establishing new roads, railways, sea links, and freight terminal. The most feasible set of infrastructure projects (i.e., actions) is selected for an efficient design of the multimodal freight transport network. The optimization process is essentially a bilevel framework in which the multimodal multiclass user equilibrium traffic assignment technique is incorporated within the lower level, while the combination of actions is approximately optimized using metaheuristics-based procedures in the upper level. The model is then applied to a large-scale transport network in the Philippines to investigate how it can contribute for improving the interregional freight transport network. Results prove that the model proposed could provide useful information in terms of the effective design of multimodal freight transport networks.

10.4.2
Metaheuristics for Hazmat Transportation

Metaheuristics have been used by several authors to deal with the problem of transporting hazardous materials. Two different optimization problems are commonly encountered in this area: the problem of transporting a single shipment from its source to its destination (essentially a variant of the *shortest-path problem*) and the problem of routing several shipments using a fleet of vehicles (a *vehicle-routing problem*).

Carotenuto, Giordani, and Ricciardelli (2007) and Carotenuto *et al.* (2007) consider a two-stage approach to solve the hazmat shipment routing and scheduling problem. In the first stage, a set of minimum and equitable risk alternative routes is generated. For each given origin–destination pair, a set of paths is selected such that those paths minimize the total risk for the population of the regional area in which the transportation network is embedded. Additionally, the risk equity is considered by trying to spread the risk among the populated links of the network. Two heuristics are proposed and tested, both based on the algorithm to find the k shortest loopless paths in a network proposed by Yen (1971). In the second stage, the different hazmat transportation requests have to be scheduled on a daily basis. Thus, a route, from the set generated previously, and a departure time must be defined, such that the sum of the hazmat shipment delays is minimized and any two hazmat vehicles are sufficiently apart from each other. The hazmat shipment scheduling problem is formulated as a special case of the job-shop scheduling model with processing alternatives (AJSP) in which each job has alternative routes, exactly one of them must be chosen for job execution, and no-wait constraints are considered. A tabu search algorithm is proposed to solve heuristically the scheduling model.

Erkut and Alp (2007) propose a methodology that can be used to identify routes for transporting hazardous materials in and through a major population centers. First, the authors consider a simplification of the hazmat network design problem in which the designed network provides only one route between any given origin and destination, i.e. the resulting network is a tree network. This problem is called the minimum-risk hazmat tree (MRTH) problem. Although the MRTH is difficult to solve optimally for large instances, several heuristics have been proposed by Erkut and Alp (2004). A tree design for a hazmat transport network may not be economically acceptable, due to the fact that that it imposes a single route for each carrier that could additionally result in excessively large and costly routes. Therefore, a greedy algorithm is used to add paths to the current network in order to allow more flexibility to the carriers. The paths to include are selected such that the increase in the total risk is minimized.

Another issue that has been considered is the multiobjective nature of the hazmat transportation problem and the fact that it involves several different stakeholders with conflicting priorities and viewpoints. Jassbi and Makvandi (2010) formulate the hazmat shipment problem as a multiobjective decision making problem. Six different objectives are considered: shortest travel distance, minimum population exposure, minimum societal risk, minimum accident probability, safest road (minimum robbery rate), and minimum terrorist attacks average. An aggregated objective function is defined as the weighted linear sum of the objectives. The approach uses a genetic algorithm to solve the optimization problem given a set of weights. The framework can use different methods to determine the weights, but the model additionally allows the decision maker to revise priorities and the resulting weights based on user feedback.

However, the more realistic case, in which several objectives are considered when a fleet of vehicles have to serve a predefined set of customers, instead of the single origin–destination case, has received less attention than the single origin–destination case. Tarantilis and Kiranoudis (2001) propose a model for the biobjective vehicle-routing problem with time windows in the context of transporting hazardous materials. The objectives taken into account are the minimization of the total distance traveled by the trucks and the minimization of the number of people placed at risk (population exposure risk). The problem is tackled by using a list-based threshold accepting algorithm (LBTA). This algorithm belongs to the class of threshold metaheuristics that have the advantage of requiring only a single parameter to be tuned. A study case is presented, that deals with the transportation of gas cylinders on the road network of Greater Athens area, involving around 10 000 roads.

Zografos and Androutsopoulos (2004) present a biobjective model considering the total travel time of the distribution process and the transportation risk as the objectives. The biobjective model is transformed into a single-objective model using a weighting approach, and both routing and path choices are performed based on this single-objective transformed model. The authors propose an algorithm to tackle the routing problem that is essentially an insertion algorithm, i.e. it builds the routes step by step by inserting in the already existing routes a new demand point at each iteration. This model and heuristic algorithm have been used by Zografos and Androutsopoulos (2008) to develop a geographical information system (GIS)-based decision support tool for integrated hazardous-materials routing and emergency-response decisions.

Giordani, Taniguchi, and Yamada (2010) consider a related problem in which three objectives are involved: total number of vehicles in use, total scheduling time of all the vehicles in operation, and total risk exposure associated with the transportation process. The problem is solved by an algorithm based on ant-colony optimization. Test results show that the algorithm performs exceptionally well on small instances, while for large instances the results are satisfactory although not comparable with the results obtained with exact approaches.

10.5 A Metaheuristic for Multimodal Hazmat Transportation

The previous sections may have been somewhat illusive for people unacquainted with optimization in general and metaheuristics in particular. In this section, we therefore describe how a (simple) metaheuristic for a (simple) hazmat transportation problem might function. We will not implement and/or test the metaheuristic–that would be a chapter in itself–but rather describe the rationale behind the design choices made at different points. The problem is the following: given a network of transfer nodes, connected by one or more transport links, find the best possible way to transport a certain quantity of goods from a node A to a node B. In this hypothetical example node A represents the production facility, node B the customer and all the other nodes are transfer stations, i.e. intermodal terminals where the product can be transshipped.

Each transport link represents the transport of goods from one transfer node to the next, and has associated with it a measure of the "distance" d (e.g., the time) and a measure of the "risk" r (the risk involved in transporting across this node). Transfer nodes may be connected by more than one link if several transport modes link the intermodal terminals (e.g., the distance between two pipeline stations can also be covered by truck). An example of such a network can be found in Figure 10.4. Note that the network contains no loops, so that each transport link takes the goods farther from A and closer to B.

This problem has two objectives: (i) the total distance $D(x)$, which is the sum of all individual distances on the path from A to B, and (ii) the total risk $R(x)$, the sum of all individual risk factors on the path from A to B.

The metaheuristic we describe in this section is an example of a GRASP metaheuristic, a simple constructive metaheuristic that combines greediness and randomness in the search for good solutions. Since the problem is multiobjective, the goal of our metaheuristic is to find a set of mutually nondominated solutions.

Before the actual start of the metaheuristic, a set of weights for the objectives and are selected. A solution x can now obtain a single weighted objective score . The same goes for each link in the network, i.e. each transport link l can be

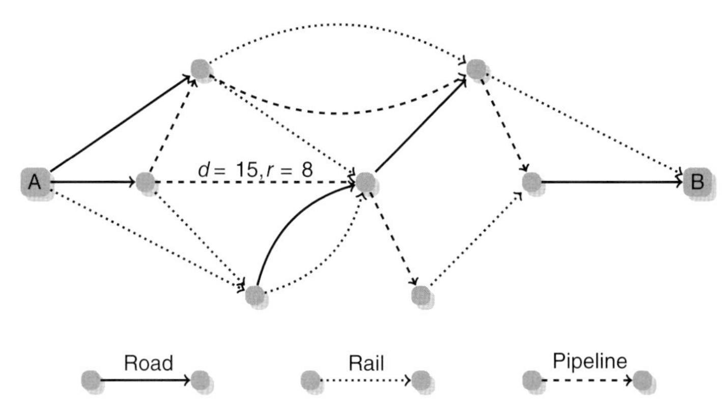

Figure 10.4 An example intermodal network: links indicate possible transportation via a particular transport mode; the distance and safety coefficients for one link are indicated as an example.

assigned a single weighted score $w_l = d_l w_d + r_l w_r$ where d_l is the distance of transport link l, and r_l is its risk.

The metaheuristic is an adaptation of a (greedy) *nearest-neighbor* heuristic, that starts from the origin, and selects at each iteration the "closest" node. The drawback of such an algorithm is that it only generates a single solution that is – in all likeliness – not optimal. Therefore, the GRASP metaheuristic randomizes this process.

To this end, the GRASP metaheuristic starts from node A and creates a list of links to nodes that can be reached from A. This list is sorted in decreasing order of w_l and only the α smallest links are retained on the list. The result is called the *restricted candidate list*. Crucially, the link to move on is selected *randomly* from the restricted candidate list.

α is a parameter of the algorithm that determines the "greediness". If $\alpha = 1$, the algorithm is exactly the same as the nearest neighbor heuristic. If α is equal to the number of nodes that can be reached from A, then the heuristic becomes a random heuristic.

When a link has been selected from the restricted candidate list, the metaheuristic moves to the next node and repeats the previous steps: a new restricted candidate list is created and a random link is selected from this link. This process continues until node B is reached.

Reaching B completes a single solution, but the GRASP metaheuristic can be repeated to create multiple, different solutions, a result of the randomness introduced in the selection from the restricted candidate list. We can, e.g., generate 100 solutions by repeating the GRASP metaheuristic 100 times.

Each time a solution is generated, the metaheuristic checks whether it is dominated by one or more solutions previously generated. If this is not the case, i.e. the solution is nondominated, it is added to an *archive*. Solutions in the archive that are dominated by the new solution, are removed. The final result of the

GRASP metaheuristic is an archive of mutually nondominated solutions that can be presented to the transport planner.

Note that this metaheuristic is about as simple as a metaheuristic can be, and that many possible advanced features have been deliberately left unmentioned for reasons of clarity. For example, adaptive selection of objective weights, local search improvement of a solution generated by the GRASP metaheuristic, clever archiving mechanisms, as well as many others, can be added to improve the performance of the method.

10.6 Conclusions and Research Opportunities

Multimodal hazmat transportation presents a number of organizational challenges, that can be solved by modeling these problems as optimization problems and solving them using an appropriate optimization technique. In this chapter we have presented a broad (and therefore necessarily shallow) overview of the current scientific consensus view of the field of metaheuristics. We have motivated why this broad class of optimization techniques presents the best choice to solve the complex problems that arise in multimodal hazmat transportation optimization problems. Metaheuristics are the methods of choice for solving large and complicated problems in a wide variety of domains closely related to multimodal hazmat transportation, and have shown in those fields that they can be put to effective use in a wide variety of practical situations. Metaheuristics can be used to solve operational problems in a few seconds, but are also able to tackle tactical and strategic problems, that are more complicated but for which computing-time requirements are not so strict.

Effectively modeling and solving large and complicated multimodal hazmat transportation problems remains a formidable challenge, however. This is exemplified by the lack of scientific literature on this topic. Although several contributions have been published on the use of metaheuristics to solve hazmat transportation problems, as well as on metaheuristics to solve multimodal transportation problems, the intersection of both sets remains remarkably empty. The development of effective metaheuristics for the multimodal transportation of hazardous materials therefore remains a considerable challenge for future research.

Bibliography

Ayed, H., Habbas, Z., and Khadraoui, D. (2009) *ACO for solving a multimodal transport problems using a transfer graph model*. International conference on computers and industrial engineering.

Ayed, H., Galvez-Fernandez, C., Habbas, Z., and Khadraoui, D. (2010) Solving time-dependent multimodal transport problems using a transfer graph model. *Computers & Industrial Engineering*. doi: 10.1016/j.cie.2010.05.018.

Beyer, H.G. and Schwefel, H.P. (2002) Evolution strategies – A comprehensive introduction. *Natural Computing*, **1** (1), 3–52.

Carotenuto, P., Giordani, S., and Ricciardelli, S. (2007) Finding minimum and equitable risk routes for hazmat shipments. *Computers & Operations Research*, **34** (5), 1304–1327. Hazardous Materials Transportation.

Carotenuto, P., Giordani, S., Ricciardelli, S., and Rismondo, S. (2007) A tabu search approach for scheduling hazmat shipments. *Computers & Operations Research*, **34** (5), 1328–1350. Hazardous Materials Transportation.

Castelli, L., Pesenti, R., and Ukovich, W. (2004) Scheduling multimodal transportation systems. *European Journal of Operational Research*, **155** (3), 603–615.

Deb, K., Pratap, A., Agarwal, S., and Meyarivan, T. (2002) A fast and elitist multiobjective genetic algorithm: NSGA-II. *IEEE Transactions on Evolutionary Computation*, **6** (2), 182–197.

Dorigo, M., Maniezzo, V., and Colorni, A. (1996) Ant system: optimization by a colony of cooperating agents. *IEEE Transactions on Systems, Man, and Cybernetics, Part B: Cybernetics*, **26** (1), 29–41.

Dorigo, M., Birattari, M., and Stützle, T. (2006) Ant colony optimization. *IEEE Computational Intelligence Magazine*, **1** (4), 28–39.

Erkut, E. and Alp, O. (2004) *A comparison of heuristics for the hazmat network design problem*. Research Report 2004-1, Department of Finance and Management Science, University of Alberta School of Business, Edmonton, Alberta, Canada.

Erkut, E. and Alp, O. (2007) Designing a road network for hazardous materials shipments. *Computers & Operations Research*, **34** (5), 1389–1405. Hazardous Materials Transportation.

Erkut, E. and Verter, V. (1998) Modeling of transport risk for hazardous materials. *Operations Research*, **46** (5), 625–642.

Erkut, E., Tjandra, S.A., and Verter, V. (2007) Hazardous materials transportation. *Handbooks in Operations Research and Management Science*, **14**, 539–621.

European Commission (2006) *Keep Europe moving – Sustainable mobility for our continent – Mid-term review of the European Commission's 2001 transport white paper.* Communication from the Commission to the Council and the European Parliament COM/2006/0314 final.

Feo, T.A. and Resende, M.G.C. (1995) Greedy randomized adaptive search procedures. *Journal of Global Optimization*, **6** (2), 109–133.

Galvez-Fernandez, C., Khadraoui, D., Ayed, H., Habbas, Z., and Alba, E. (2009) Distributed approach for solving time-dependent problems in multimodal transport networks. *Computers & Industrial Engineering*, **61** (2), 319–401.

Giordani, S., Taniguchi, E., and Yamada, T. (2010) Ant colony system based routing and scheduling for hazardous material transportation. *Procedia – Social and Behavioral Sciences*, **2** (3), 6097–6108. The Sixth International Conference on City Logistics.

Glover, F. (1977) Heuristics for integer programming using surrogate constraints. *Decision Sciences*, **8** (1), 156–166.

Glover, F. (1986) Future paths for integer programming and links to artificial intelligence. *Computers & Operations Research*, **13**, 533–549.

Glover, F. (1989) Tabu search-part I. *ORSA Journal on Computing*, **1** (3), 190–206.

Glover, F. (1990) Tabu search-part II. *ORSA Journal on Computing*, **2** (1), 4–32.

Glover, F. and Laguna, M. (1993) Tabu search, in *Modern Heuristic Techniques for Combinatorial Problems* (ed. C.R. Reeves), Halstead Press, Australia, pp. 70–141.

Goldberg, D.E., *et al.* (1989) *Genetic Algorithms in Search, Optimization, and Machine Learning*, Addison-Wesley, Reading, Menlo Park.

Holland, J.H. (1975) *Adaptation in Natural and Artificial Systems*, University of Michigan Press, Ann Arbor.

Jansen, B., Swinkels, P.C.J., Teeuwen, G.J.A., van Antwerpen de Fluiter, B., and Fleuren, H.A. (2004) Operational planning of a large-scale multi-modal transportation system. *European Journal of Operational Research*, **156** (1), 41–53.

Jassbi, J. and Makvandi, P. (2010) Route selection based on soft modem framework in transportation of hazardous materials. *Applied Mathematical Sciences*, **4** (63), 3121–313261, http://www.or.journal.

informs.org/cgi/content/abstract/36/1/84 (accessed September 1, 2011).

Jones, D.F., Mirrazavi, S.K., and Tamiz, M. (2002) Multi-objective meta-heuristics: an overview of the current state-of-the-art. *European Journal of Operational Research*, **137** (1), 1–9.

Kirkpatrick, S., Gelatt, C.D. Jr., and Vecchi, M.P. (1983) Optimization by simulated annealing. *Science*, **220** (4598), 671.

Knoppers, P. and Muller, T. (1995) Optimized transfer opportunities in public transport. *Transportation Science*, **29** (1), 101–105.

Knowles, J. and Corne, D. (1999) The Pareto archived evolution strategy: a new baseline algorithm for Pareto multiobjective optimisation, in *Proceedings of the 1999 Congress on Evolutionary Computation (CEC '99)*, pages 98–105.

Koza, J.R. (1992) *Genetic Programming: on the Programming of Computers by Means of Natural Selection*, The MIT press.

Lourenco, H., Martin, O., and Stützle, T. (2003) Iterated local search, in *Handbook of Metaheuristics* (eds F.W. Glover and G.A. Kochenberger), Springer, pp. 320–353.

Marler, R.T. and Arora, J.S. (2004) Survey of multi-objective optimization methods for engineering. *Structural and Multidisciplinary Optimization*, **26** (6), 369–395.

Michalewicz, Z. and Fogel, D.B. (2004) *How to Solve It: Modern Heuristics*, Springer-Verlag, New York

Moscato, P. (1989) On evolution, search, optimization, genetic algorithms and martial arts: towards memetic algorithms. *Caltech Concurrent Computation Program, C3P Report*, **826**, 1989.

Nagy, G. and Salhi, S. (2007) Location-routing: issues, models and methods. *European Journal of Operational Research*, **177** (2), 649–672.

Pradhananga, R., Taniguchi, E., and Yamada, T. (2010) Ant colony system based routing and scheduling for hazardous material transportation. *Procedia – Social and Behavioral Sciences*, **2** (3), 6097–6108.

Rondinelli, D. and Berry, M. (2000) Multimodal transportation, logistics, and the environment: managing interactions in a global economy. *European Management Journal*, **18** (4), 398–410.

Sörensen, K. and Glover, F. (to appear) Metaheuristics, in *Encyclopedia of Operations Research and Management Science* (eds S.I. Gass and M. Fu), Springer, New York.

Sörensen, K., Sevaux, M., and Schittekat, P. (2008) "Multiple neighbourhood search" in commercial VRP packages: evolving towards self-adaptive methods, in *Lecture Notes in Economics and Mathematical Systems*, vol. 136 (eds C. Cotta, M. Sevaux, and K. Sörensen), chapter Adaptive, self-adaptive and multi-level metaheuristics, Springer, London., pp. 239–253.

Sörensen, K., Vanovermeire, C., and Busschaert, S. (2011) *Efficient metaheuristics for the intermodal terminal location problem*. Working Paper 2011/02, University of Antwerp, Faculty of Applied Economics.

Talbi, E.G. (2009) *Metaheuristics: from Design to Implementation*, John Wiley & Sons, Inc.

Tarantilis, C.D. and Kiranoudis, C. (2001) Using the vehicle routing problem for the transportation of hazardous materials. *Operations Research*, **1**, 67–78. 10.1007/BF02936400.

The Economist. *Riders on a swarm*, http://www.economist.com/node/16789226 (accessed 12 August 2010).

Verma, M. (2009) A cost and expected consequence approach to planning and managing railroad transportation of hazardous materials. *Transportation Research Part D: Transport and Environment*, **14** (5), 300–308.

Verma, M. and Verter, V. (2007) Railroad transportation of dangerous goods: population exposure to airborne toxins. *Computers & Operations Research*, **34** (5), 1287–1303.

Walshaw, C. (2004) Multilevel refinement for combinatorial optimisation problems. *Annals of Operations Research*, **131** (1), 325–372.

Yamada, T., Russ, B.F., Castro, J., and Taniguchi, E. (2009) Designing multimodal freight transport networks: a heuristic approach and applications. *Transportation Science*, **43** (2), 129–143.

Yen, J.-Y. (1971) Finding the K shortest loopless paths in a network. *Management Science*, **17** (11), 712–716.

Zhang, J., Hodgson, J., and Erkut, E. (2000) Using GIS to assess the risks of hazardous materials transport in networks. *European Journal of Operational Research*, **121** (2), 316–329.

Zitzler, E., Laumanns, M., Thiele, L., Zitzler, E., Zitzler, E., Thiele, L., and Thiele, L. (2001) *SPEA2: Improving the Strength Pareto Evolutionary Algorithm*, Citeseer.

Zografos, K.G. and Androutsopoulos, K.N. (2004) A heuristic algorithm for solving hazardous materials distribution problems. *European Journal of Operational Research*, **152** (2), 507–519.

Zografos, K.G. and Androutsopoulos, K.N. (2008) A decision support system for integrated hazardous materials routing and emergency response decisions. *Transportation Research Part C: Emerging Technologies*, **16** (6), 684–703.

11
Freight Security and Livability: US Toxic and Hazardous Events from 2000 to 2010

Lisa Schweitzer, Pamela Murray-Tuite, Daniel Inloes, Jr., Mohja Rhoads, and Fynnwin Prager

11.1
Introduction

After nearly a decade of freight policy focused on security and expansion, recent US Federal policy under the Obama administration has begun to stress an entirely new direction: livability. Livability attempts to balance the needs that nearby residents have for environmental quality with the building, operations, and maintenance of nearby freight facilities. This chapter examines the consequences for nearby communities of hazardous freight, both from accidental and, by extension, terrorist events.

Freight shippers manage over 323 billion ton-miles of toxic and hazardous materials every year, and that volume has grown over time along with the US economy. Serious incidents, though, are rare. From 2000 to 2010, the US had over 120 000 spills recorded from around the country, with roughly 5000 listed as serious over that time. Loss of human life or injuries are infrequent: only 136 people have died from hazardous or toxic material exposure, while only 1587 have been injured in the last decade. Nonetheless, when accidents do become serious, they can cause considerable economic damage. The total economic damage associated with no-notice hazardous materials spill events exceed $550 million, with very serious single events that cost in excess of $20 million.

Moreover, the past decade of accident data suggest vulnerability to terrorism as well as accidents. Over 150 000 people were evacuated during the past decade because of accidental spills. The success of those evacuations hinges on the reliability of information and practitioners engaged in freight shipping–two factors that may be expected to break down under a planned, intentional strike such as a terrorist action. Under conditions where information placards cannot be trusted and where personnel or onlookers may be complicit and malicious, the consequences may be much higher.

Our past experiences with toxic and hazardous materials (hazmat) evacuations can yield insights into the consequences of terrorist strikes at or near large-scale multimodal facilities in the US. The results of evacuations conducted in

Security Aspects of Uni- and Multimodal Hazmat Transportation Systems, First Edition. Edited by Genserik L.L. Reniers, Luca Zamparini.

"best-case" accidental conditions serve as a possible lower bound for damage estimates – the optimistic case – of terrorist acts against the hazmat system and suggest what the consequences of these events may be for communities surrounding large-scale freight facilities.

11.2 Background

Prior to the industrial revolution, goods movement occurred predominately by horse, barge, and foot (Jackson, 1987; Cronon, 1992). Workers and traders flocked to housing near freight facilities and ports out of economic advantage. Many of today's megaregions began as port cities – entryways for trade activities – and, as a result, these locations have always been targets during armed conflicts and sources of environmental vulnerability. Just as people today worry about the global threat posed by viruses spread through global transportation networks, armies and goods movement spread diseases, perhaps most infamously the bubonic plagues of the 14th century (Kelly, 2006). Horse-powered cities were fetid, places where pedestrians routinely risked typhus and other pathogen-related illnesses from sharing their streets with piles of manure and the rotting corpses of horses that had been worked to death on the city's streets (McShane and Tarr, 2007).

With steam, rail and streetcar technologies, workers and traders could cover more distance in less time, opening spatial opportunities for where they could live and work relative to factories, trade centers, and warehousing (Jackson, 1987). In addition to new transport technologies, nuisance laws, and, eventually, zoning codes in the US instituted the social, economic, and geographic separation of urban housing, particularly for the affluent, from freight and industry (Ferrey, 2010). As regions have grown, so have calls to reverse the spread of urban populations through infill and higher density development and by doing away with single-use zoning that separates people from employment and trade (Bernick and Cervero, 1997; Levine, 2006).

Ultimately, the push and pull factors of policy, planning, and new technologies have had two major effects that interest us here. First, urban population growth (through natural increase and long-term, sustained outmigration from rural areas to urban centers) has placed more people than ever before into very high levels of population density within metropolitan centers. Just as an example, the Port of Los Angeles was established formally (after decades as a harbor) in 1906. The Los Angeles population was a little over 102 000 people then. Today, the city of Los Angeles, has close to 4 million people, with the surrounding metropolitan area closer to 16 million, surrounding the US's two largest freight ports. Freight shipping as an industry has grown over the past century as well, particularly over the last decades of the 20th century, as global capital flows have increased, with logistics and industry practices moving towards greater scale and scope of goods movement facilities. High volumes of materials are being moved closer to higher numbers of urban residents as a result of these two growth effects.

Federal regulations have fostered both freight consolidation and scale, particularly in hazardous-materials transport. The Resource Conservation and Recovery Act (RCRA) in 1980 mandated cradle-to-grave tracking of toxic and hazardous material as it moves through the US (Ferrey 2010). RCRA and subsequent laws requiring drivers and handlers to have additional credentials, standardized containers, and placarding, have created barriers to entry in hazmat shipping. The regulatory environment yielded predictable results: the fewer shippers and facilities, the easier it is to monitor and enforce high industry safety standards.

Nonetheless, consolidation in the hazmat freight industry can have multiple–and unfortunately conflicting–effects on community vulnerability when populations have grown up around freight facilities. Consolidation can build up the volumes and diversity of materials at one geographic location. On the one hand, the readily identifiable location helps first responders know where the likely problems are, and in the case of everyday incidents, allows companies to keep specialized equipment and professionals on site, in case something does happen. The economies of scale and scope realized during everyday shipping activities also manifest for incident response. On the other hand, the consolidation of hazardous materials freight in one geographic location creates increased risks of accidental releases and a readily identifiable location for terrorist acts.

Toxic and hazardous-materials shipping reflects the perennial tension between consolidating and distributing hazard in urban contexts: is it more secure (i.e. less likely to cause death, damage, and injury) to consolidate risks onto one location and one set of large-scale networks? Or is it more secure to disperse risks in small amounts, carried discretely through a highly disaggregated network of smaller-scale facilities and transport modes?

The concepts driving these questions are illustrated in Figure 11.1. Networks A and B illustrate the land use, infrastructure, and industrial organization that most similarly represents the arrangement of multimodal facilities in the United States.

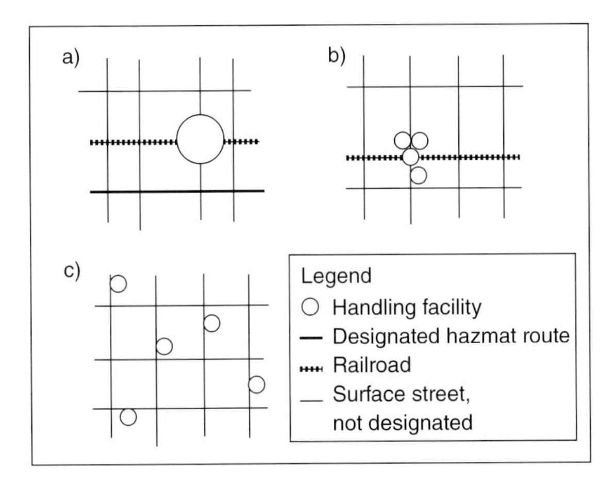

Figure 11.1 Different network, industrial organization, and land use.

Industrial consolidation can prompt companies to pursue very large operations, as in A. However, zoning and industrial agglomeration can cause the geographic clustering shown in scenario B, where multiple companies, and even multiple hazmat handling industries appear in a spatial area. In either A or B, vulnerability centers on one, specific geographic location.

A key difference between A and B, however, concerns the highway network. In both A and B, the facilities are served by only one railway, which itself poses a potential target. The network in A reflects the current state of the practice in the US, which designates specific routes, while disallowing others, for highway movements of hazardous materials. In so doing, the requirement demonstrates the benefits of managing materials for accidents: routing is done according to the highway capacity and safety standard, and isolates hazmat traffic to specific links. By disallowing other routes, however, hazmat route designation makes it easier for outsiders to figure out what highway segments are likely places for hazardous content to appear.

Scenario C shows the opposite of the three variables (land use, infrastructure, and industrial organization). The land-use configuration and industrial organization separate the facilities across the geographic network. The disaggregate, gridded highway network allows for many routing combinations, once past the limits of facility-access links. This routing flexibility allows shippers to vary routes for security purposes and/or avoid minor disruptions in the network. Scenario C lacks rail transport, which would allow hazmat shipments to be easily tracked and controlled, but rail has limited routing flexibility and the volume of materials carried at a given time has potentially disastrous effects if an accident or attack occurs.

Without information about the shipments and, more importantly, the population of the surrounding area, it is impossible to determine what type of arrangement carries the highest vulnerability for urban populations. However, the existing US conditions currently resemble A and B, and the US is unlikely to shift land uses or hazardous materials management. The result is a geographic consolidation of hazmat risks at multimodal facilities or facility clusters, and the designated routes that immediately serve those area facilities.

Theorizing about risks anchored by facilities and surrounding land uses reconceptualizes risk away from largely stochastic events – which can happen anywhere towards a more tractable geography of risk. Most studies examine risks according to routes and attempt to derive the population-minimizing routes between origins and destinations (Purdy, 1993; Erkut and Verter, 1998; Pet-Armacost *et al.*, 1999, and Leonelli, Bonvicini, and Spadoni, 1999) or a combined objective of minimizing travel time and population exposure or other measures of risk (e.g., Erkut and Verter, 1998; Nozick, List, and Turnquist, 1997; Verter and Erkut, 1997; Sherali *et al.*, 1997; Nembhard and White, 1997; Sulijoadikusumo and Nozick, 1998; Dadkar, Jones, and Nozick, 2008). The population minimizing route may not always be the shortest route or the route that uses the highest-standard facilities; hence, researchers have often included dual objectives for hazmat routing. In addition, examinations of route-based risk functions tend to treat hazmat spill likelihood as a function of distance (Samuel *et al.*, 2009), but such conceptualizations of stochastic events are less useful in thinking about the likely geographic location of intentional strikes.

Analyses of the risks for terrorist events specifically at multimodal facilities tend to be primarily focused on seaports and on the loss of economic productivity from shutting down freight facilities or critical infrastructure (Rose, 2009). For hazardous materials, the research contains mostly frameworks and many potential "how-tos." Strikes against freight facilities are unknown in the US; internationally, above-ground oil pipelines have been targeted, not to harm nearby populations but to disrupt production and send a message to corporate owners (Frynas, 1998; Apter, 2005). Thus, the available empirical knowledge base and data for building vulnerability or consequences models of intentional strikes are sparse to nonexistent.

To give some indicator of the relationship between facilities and potential consequences, we can examine the past record of accidental incidents, their geographic locations, and their consequences on communities surrounding the multimodal facilities. In this way, it is possible to test empirically whether the industrial organization and infrastructure networks laid out conceptually in Figure 11.1 concentrate accidental hazardous-materials shipping risks in ways that can help enlighten the potential consequences of terrorist strikes. Moreover, the consequences from accidental spills provide further information for future risk-modeling efforts.

11.3 Data on Consequences

There is a fairly comprehensive set of records available in the US for examining the spill records surrounding multimodal facilities, although the data have some problems with geographic accuracy, particularly for data going back further than 2000.

11.3.1 National Transportation Atlas Database 2010

We define multimodal facilities as those listed in the Atlas database, published by the Bureau of Transportation Statistics (BTS). These data contain the name of the facility, city, state, zip code, list of businesses associated with the facility, and mode. According to the Atlas, there are 3281 intermodal facilities in the US: 227 rail–truck-facilities, 744 port–rail–truck facilities, 408 air–truck facilities, 62 port–truck facilities, 10 rail–port, and one port–rail–truck–airport (Port of Little Rock).

11.3.2 Hazardous Materials Information Reporting System (HMIRS)

The HMIRS data are compiled from reports made by transporters and first responders. The data include many variables regarding the incident: the time of day, materials spilled, amounts, the carriers, and what triggered the incident (e.g., human error, container failure). The data also include information on the consequences of the spills, including the number of people killed, injured, or evacuated. The data are collected, maintained, and distributed by the U.S. Department of Transportation's Office of Hazardous Materials Safety (OHMS). The data used for our analysis project span from 2000 to 2010. The OHMS designated serious spills as those that cause death or injury, close a major road, or prompt an evacuation of more than six people.

11.3.3 Hazardous Substances Emergency Events Surveillance (HSEES)

The HSEES data system is maintained by the U.S. Department of Health and Human Services Agency for the Toxic Substances and Disease Registry. The program includes cooperative reporting agreements with the state health departments of 14 states, including New York, Florida, New Jersey, Texas, and Louisiana. The HSEES is a public database that records information on acute hazardous materials releases and their consequences. It has been in existence since 1990. The departments of health at the state level report the geographic location, timing, substances, and volumes of the release, and release consequences, such as evacuations, injuries, and fatalities. These spills are also listed in the HMIRS data, but they may have additional details regarding the long-term effects of the spills.

11.3.4 Lexis-Nexis and Newspaper Reporting on Serious Spill Incidents

In order to expand the information in the HMIRS and the HSEES, the research team used Lexis-Nexis to find newspaper coverage of the major spill incidents. A member of the research team cross-referenced Lexis-Nexis against serious spills in the HMIRS database by date, location, and substance (three separate searches). The match rate was disappointing. We found news coverage of only 22 per cent of the serious spills that occurred across the US, and of those, only 15 per cent related to spills occurring during transfer or storage at multimodal facilities. However, the records were saved for which events did receive press coverage. In some respects, the lack of press coverage demonstrates how well hazmat materials incidents are managed; however, it also demonstrates how invisible hazardous-materials shipping is to the general populace.

11.4 Consequences and Geography

We restrict the spatial analysis to the state of California because of the computational requirements of doing a spatial analysis on the entire country, given the 120 000 spills that occurred from 2000 to 2010. Instead, 10 486 spills occurred in

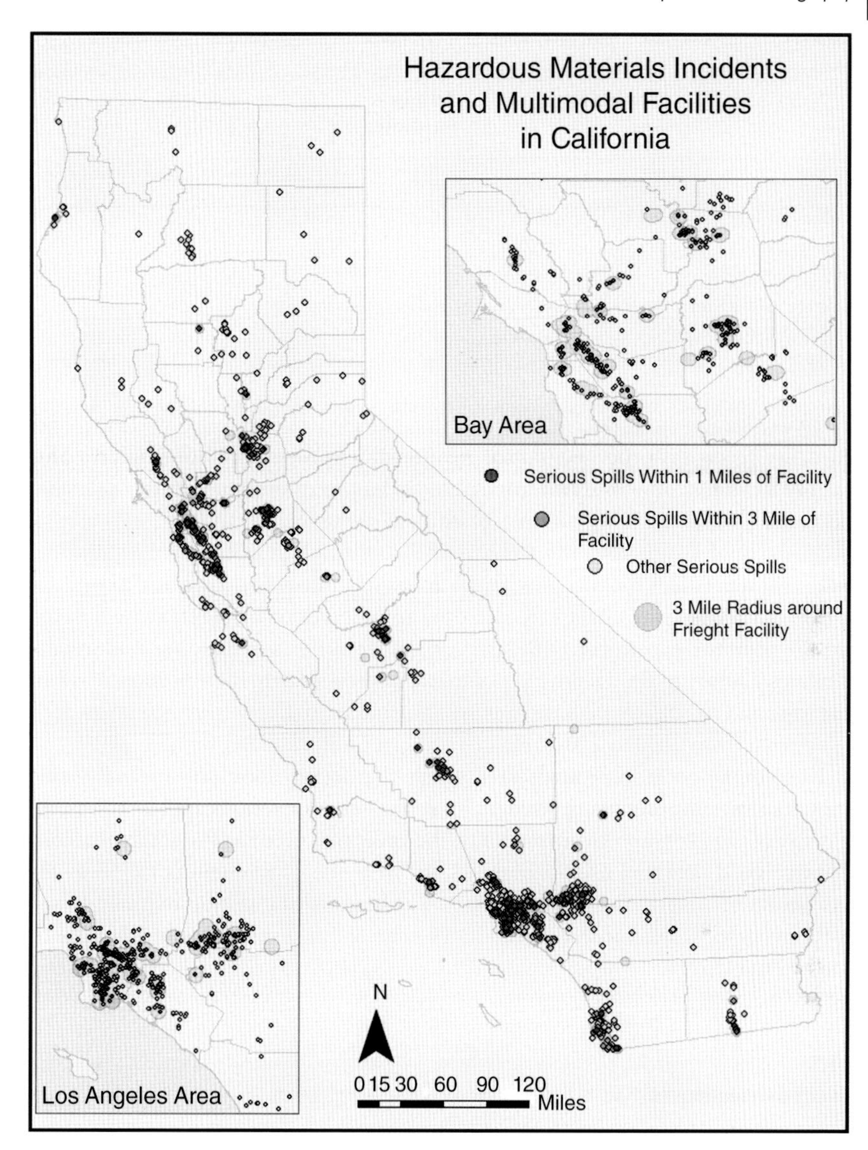

Figure 11.2 Spatial buffers around multimodal facilities in California.

California, which makes for a much more tractable spatial analysis. The data were geocoded to a 91 per cent match. All of the multimodal facility locations already had geographic location information. The California spills were mapped against facility location, with the results shown in Figure 11.2, with map insets for the Los Angeles and Bay Area regions.

Figure 11.2 only maps the serious spills against the multimodal facility location: with all spills, there was too much overplotting to distinguish the relationships.

Table 11.1 Spills within 1 mile or 3 miles of multimodal facilities, 2000 to 2010.

	Total incidents		Serious incidents	
Total	10 486	100%	1109	100%
1-mile buffers	3 393	32%	314	28%
3-mile buffers	6 531	62%	631	57%

Source: HMIRS and National Transportation Atlas data, geocoded and analyzed by the authors.

In order to capture the geographic relationship between spill and multimodal facility locations, one-mile and three-mile buffers were used to capture spills that occurred on highway and rail links proximate to the facilities.

Spatial analysis of the buffers shows that a third of all spills occur within one mile and more than half occur within 3 miles of multimodal facilities, as seen in Table 11.1. These percentages are mirrored among serious incidents as well. Based on the previous experience with spills, a spatial buffer surrounding multimodal facilities capturing the facility-access link captures a fairly high portion of all of the toxic and hazardous materials spills. This finding tracks with previous research conducted from 2000 to 2004 only in southern California (Schweitzer, 2006).

This simple geographic analysis suggests that facility locations are reasonable spatial proxies for predicting accident locations – and for serious spill locations. As a result of the geographic commonality, the spills that occur in the accident record are also good potential exemplar events for what the consequences might be for strikes against the multimodal facilities. Further analysis will be required to see if the geographic relationship found in the state of California holds in other places around the US.

11.5 Event Consequences

The HMIRS data have multiple measures for event consequences that are summarized in Table 11.2. The first set of outcome variables is binary, and it indicates whether, once a spill has occurred, any subsequent event then occurred. Release measures whether, in the case of a container breech, the material leaves the container and enters the environment. Just because a vehicle with hazardous materials derails or crashes does not mean that the container will breech, and just because the container breeches does not mean that the material will always release. Of the 5196 serious spills in the US (and 120 000 total spills), 4579 spills had a release occur.

Table 11.2 Outcomes associated with toxic and hazardous events.

Outcome	Data definition
Binary variables (1 = Yes, 0 = No)	
Release (spillage) (*r*)	Material is released due to incident; ($N = 4579$)
Radioactive material	Release of radioactive material (extremely rare)
Closure	Major artery was closed as a result of spill; $Y = 1204$
Environmental damage	Release resulted in environmental damage; $Y = 606$
Evacuation	Release resulted in an evacuation order; $Y = 843$
Gas dispersion	Materials released in gaseous form; $Y = 687$
Fire	Material caught on fire; $Y = 472$
Explosion	Whether explosion occurred; $Y = 145$
Cost variables ($US)	
Property damage	Damage to public or private property
Response costs	Costs of labor and equipment for responders
Remediation and cleanup	Remediation costs
Total damages	Total cost figure (sum of property, response, remediation and other costs)
Continuous or count variables (persons, hours)	
Volume released	Volume of materials released
Fatalities	Fatalities associated with employees, the public, and first responders
Injuries	Injuries associated with employees, the public, and first responders
Total evacuated	Employees and public evacuated
Total evacuation hours	Duration of the evacuation
Duration of closure	Duration of major artery closure
Calculated Variables	
Value of life and injury	Deaths multiplied by the statistical value of life
Person-hours of evacuation	Duration of evacuation multiplied by members of the public evacuated.
Lost productivity	Value of time lost to evacuation: person-hours of evacuation multiplied by prevailing wage rate

Source: Hazardous Materials Information Reporting System, Codebook. Person-hours of evacuation are not reported; these are compiled by the authors.

The radioactive material (RAM) binary variable represents only one of a possible series of binary outcome variables based on the type of materials released. Radioactive material shipping is rare compared to other types of shipping, and the containers in which they are shipped are carefully constructed. Only four RAM events occurred from 2000 to 2010. The Hazardous Materials Information Systems data contain categories for all the standard classes of hazardous materials, and thus it is possible to create binary event outcomes for any type of material.

The next six binary outcome variables concern events that may occur subsequent to an incident and a release. Closure (n = 1204) measures whether the incident closed a major arterial (or higher level of service) roadway. Environmental damage (n = 606) indicates whether the spill caused any environmental damage, such as a petroleum spill. Unfortunately, the databases contain very little information about the nature of environmental damages. Evacuation (n = 843), gas dispersion (n = 687), fire (n = 472), and explosion (n = 145) represent progressively rarer events, so that the probability of any given outcome is related to the outcomes of previous events: $p(c|r)|p(r|e)$, where $p(r|e)$, the probability of a release (r), is conditional on a previous event (e) such as an intentional strike, crash, turnover, or cargo mishandle, and where the probability of any given consequence (c) is again conditioned on a release event.

Figure 11.3 shows the breakdown of the event types by mode and hazardous materials class code for the two binary outcome measures that show the most variation by mode and hazardous materials class. The plots first show that most serious spills occur for Class 3 hazardous materials, which are flammable liquids – gasoline and other fuels – as we would expect due to the prevalence of the materials.

The serious spills are distributed among hazardous materials classes similar to the prevalence of their shipping, with one exception. Corrosive materials (Class 8) are somewhat more represented in serious spills than in the entire spills record. Because there are so few spills from water transport, those are not illustrated. Infrequent hazardous materials classes are also omitted from the figures.

A contrast of the two mosaic plots shows that rail and air modes have caused proportionately more evacuation events than highway shipping for both flammable (Class 3) and corrosive materials (Class 8). This result is likely due to the volumes that can be transported by these modes, relative to a single truck. The reverse is true for events causing environmental damage, which could be a result of separation of rail and air facilities from other land uses. This separation contrasts with highways, which are more geographically dispersed and come in closer contact with environmentally sensitive areas. Here again, however, corrosive materials are proportionately over-represented among serious spills.

The binary event variables also have analogs in the HMIRS for continuous measures that reflect the extent or costs of those outcomes. The cost variables measure three significant consequences: the costs to property, the costs associated with time and equipment needed by responders to act on the event, and the remediation and clean-up costs. The data do not include costs associated with productivity loss due to closures or evacuations. Figure 11.4 displays response, remediation,

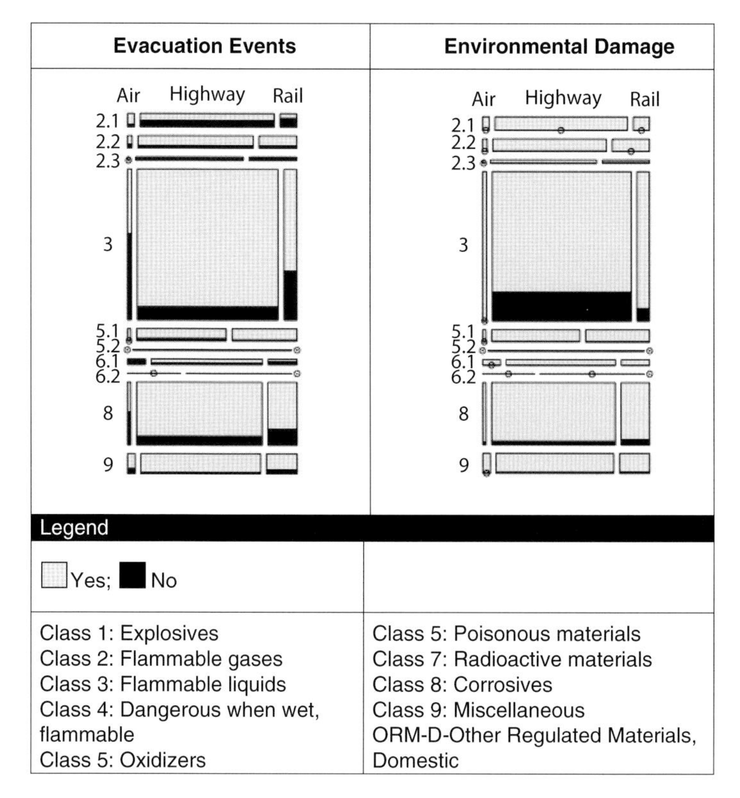

Figure 11.3 Evacuation and environmental damage by mode and hazardous materials class, 2000 to 2010. Source: Data from the Hazardous Materials Information, 2000 to 2010 compiled by the authors.

and property damage costs plotted against the total costs associated with the spill. The data points are broken out by mode symbolically. Most serious spills cause less than $5 million in damages, and all of the serious water and air hazmat events fall into that cluster of points centered at $5 million and under.

The interesting data points here are the extreme values for highway and railway, both of which had a handful of spills from 2000 to 2010 that imposed higher cost consequences than did other serious spills. Although there are only a few, scattered extreme events, rail modal events are again disproportionately represented among the cost figures. However, the most extreme consequences for response, property, and remediation costs occurred on highways. While serious railway spills were likely to prompt evacuations, highway events have imposed the highest out-of-pocket cost consequences for the companies involved in the spills. Each of these outliers may be good exemplar events for use in analyzing terrorist risks.

Other measures of consequences, such as death, injury, and time loss to evacuations can also be monetized. Since injuries and fatalities are usually monetized by a standard amount, those measures are perhaps less interesting for illustration

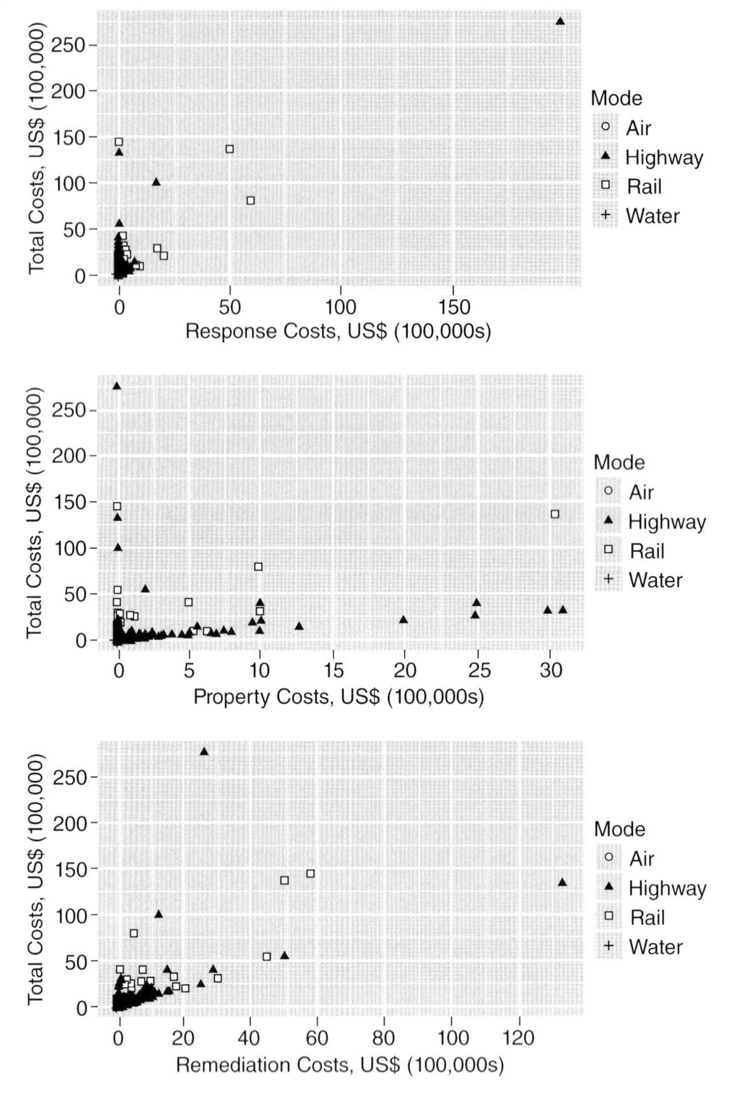

Figure 11.4 Cost consequences of serious spills by mode, 2000 to 2010. Source: Data from the Hazardous Materials Information, 2000 to 2010 compiled by the authors.

than the value of time lost due to evacuations. The value of time lost to evacuations is a function of the total number of people evacuated, the duration of the evacuation, and the value of time assigned to them.

Figure 11.5 plots the total damage costs (logged) against the person hours of time loss, again using the most prevalent hazards classes and modes. Wages or time values would be a constant, and thus they are not included here; we have allowed for the zero values to be included (modified for the log) so that the figure

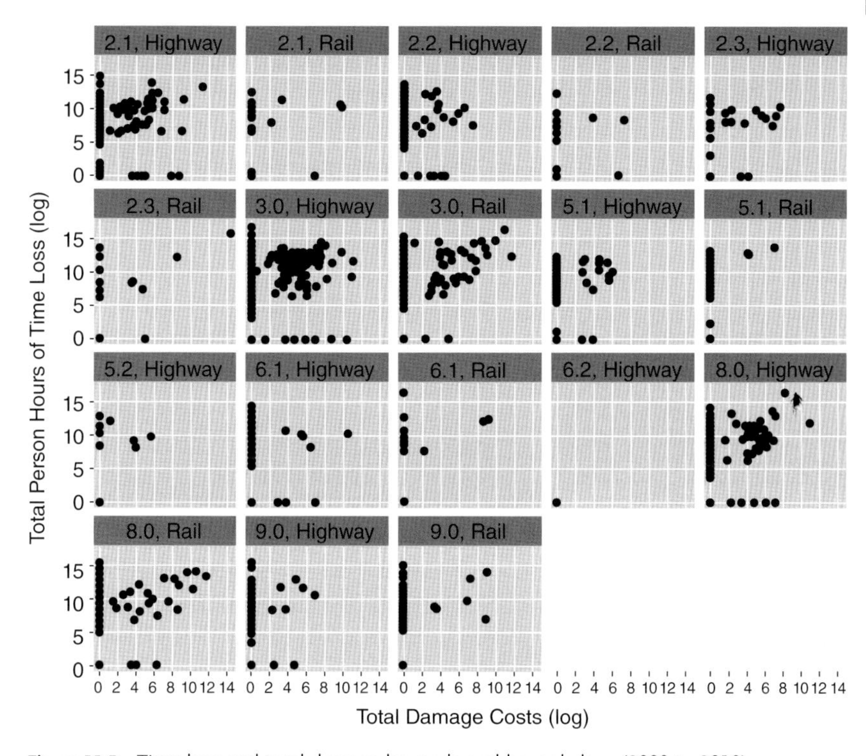

Figure 11.5 Time loss and total damage by mode and hazard class (2000 to 2010).

displays the split in the events between those that cause damage without evacuation, evacuation with low damage costs, and the third group: those events that cause both. That middle group demonstrates a strong and positive association between time loss costs and total damage costs. It is once again possible to see how three classes of materials drive the consequences for serious spills across modes: flammable gases (Class 2.1), flammable liquids (Class 3), and corrosives (Class 8). Poisonous gases transported via highway (Class 2.3) have caused more events with both evacuations and total material damage than on railways, and the same is true of oxidizers and organic peroxides (Classes 5.1 and 5.2). Because these are not spill rates, there is no information about what is more prevalent or more likely to spill. Instead, the data simply reflect the consequences of what has occurred by mode and class – not the likelihoods associated with those consequences by either mode or class.

11.6 Conclusions

Looking at the data across the US and in California shows that while most hazardous materials events are minor, there are a large number of events – roughly 10 000

Table 11.3 Data summary consequences, 2000 to 2010.

	California	CA per cent	CA share of US	US	US per cent
Tons shipped[a]	1 997 550 000	100%	9%	22 311 330 000	100%
Total events	10 626		9%	121 405	100%
Serious events	297		6%	5196	4%
Deaths (total)	3		2%	136	0%
Worst	1			9	0%
Mean	–			–	
Injuries	37			1587	
Worst[b]	5	14%	0%	631	12%
Mean	–			–	
Total evacuation	6196			154 616	
Worst[b]	2 000	32%	1%	25 000	16%
Mean	21			30	
Total evacuation (h)	429			7230	
Worst[b]	110	26%	2%	2 016	28%
Mean	1			1	
Total person-hours	135 336			2 715 356	
Worst[b]	120 000	89%	4%	1 625 000	60%
Mean	455			522	
Total property	$1 643 317			$68 748 792	
Worst[b]	490 000	30%	1%	3 100 000	5%
Mean	5533			13 230	
Total response	$2 373 122			$67 681 719	
Worst[b]	1 970 065	83%	3%	19 790 000	29%
Mean	79 903			13 030	
Total remediation	$31 069 089			$230 095 379	
Worst[b]	13 300 000	43%	6%	13 300 000	6%
Mean	104 610			44 280	
Total cost ($)	$67 738 646			$571 114 173	
Worst[b]	27 467 818	41%	5%	27 470 000	5%
Mean	228 076			12 300	

a) These data are from the US Commodity Flow Survey, 2007; other years estimated by the authors.
b) The worst-case percentages are calculated as a percentage of the US worst cases rather than all spills or all serious spills.
Source: Hazardous Materials Information System, data compiled by the authors.

every year. The past spill record for the totals and worst events are compiled in Table 11.3. The takeaway message from this table suggests that hazardous and toxic events are managed conservatively: the US definition of a "serious" spill has a fairly low threshold for damage and off-site consequences. The result is that 1 in every 23 spills in the US is considered to be serious. Of those spills, some become quite serious indeed, both in terms of evacuation costs and total damages, and as we have shown, those two figures tend to move together in a subset of all serious incidents. The relatively low numbers of lives lost and injuries attest to how well most incidents are managed. Nonetheless, the worst events, infrequent though they are, are quite serious for surrounding communities.

Given the geographic analysis from Sections 11.4 and 11.5, we established that these events are concentrated together with multimodal facilities. This clustering occurs either as the result of handling at that facility or from multimodal facility colocation with originating or destination locations through the industrial clustering within US regions. The geographic vulnerability of these locations is therefore apparent, as are the potential consequences for their residential neighbors.

For livability and vulnerability, a complex picture emerges. Multimodal freight shippers are, even with all their spills, fairly good neighbors – except for those infrequent times when an event spirals. The evidence for the livability argument – that freight and residential populations can coexist – is mixed. The consequences for human life and injuries of accidental releases have been low, especially compared to the risks and mortality resulting from passenger transport. Nonetheless, the volumes handled at multimodal facilities and the highways and railways that run through US regions are large, and a few selected events become very serious indeed.

Some exemplary events can help in further understanding the issues raised throughout this analysis. Table 11.4 shows a sample of the highest consequence

Table 11.4 A sample of exemplar events, 2000 to 2010

Location	Exemplar	Measure	Date	Route	Mode	Substance	Event
Detroit, MI	Property damage	3 100 000.00	10/6/03	I-75 Ramp	Highway	Gasoline	Cargo tank release that then caught fire
	Total damage costs	27 467 818.00					
Burbank, CA	Response costs	19 790 065.00	6/10/10	Highway 134E	Highway	Gasoline	Cargo tank turned over
Keys, CA	Remediation costs	13 300 000.00	1/27/06	Unreported	Highway	Formic acid	Tank cracked during crash
Graniteville, SC	Evacuation duration	301	1/6/05	Milepost 178.3	Rail	Chlorine	Multicar derailment
	Person-hours	1 625 400					
	Fatalities	9					
	Injuries	631					

events from around the US. Note that these do not necessarily occur at multimodal facilities, but they do serve as exemplars for events that have caused fairly serious consequences for those living near hazardous-materials shipping. Further study of these events in future research can help analysts envision the consequences of a terrorist strike. For now, they serve to illustrate a final point about security and hazmat shipping.

Comparatively common substances have had demonstrably high consequences in isolated events in the past decade. As bad as the nightmare scenario – an intentional strike against radioactive material – would be, everyday materials transport, like gasoline shipments, have prompted two of the four worst events over the past 10 years in terms of property damage and total costs. Gasoline is virtually everywhere in the US: the shipments are ubiquitous, as are gas stations. The other substance, chlorine, is also common; it has many uses in industry and government, including water treatment. These are not, in other words, exotic or infrequently handled materials. It is unlikely that the large amounts of gasoline or chlorine – or the other commonly used hazardous materials handled throughout the US every day – will decline any time soon. They provide ready and available material for terrorists to use, and those consequences may be worse than these accidental releases – which are bad enough.

It may be, therefore, a mistake to plan for strikes against multimodal facilities only in terms of highly toxic or radioactive materials. As dangerous as those substances are, they may be less readily found than other substances, and they may be isolated more from potential victims. As the US tries to move towards a livable freight agenda, these types of security issues should be analyzed in regions that have human settlements surrounding freight activities.

Bibliography

Apter, A.H. (2005) *The Pan-African Nation: Oil and the Spectacle of Culture in Nigeria*, illustrated ed., University of Chicago Press, Chicago.

Bernick, M.S. and Cervero, R.B. (1997) *Transit Villages in the 21st Century*, McGraw-Hill, New York.

Cronon, W. (1992) *Nature's Metropolis: Chicago and the Great West*, reprint, illustrated edn, W.W. Norton, New York.

Dadkar, Y., Jones, D., and Nozick, L. (2008) Identifying geographically diverse routes for the transportation of hazardous materials. *Transportation Research, Part E*, **44**, 333–349.

Erkut, E. and Verter, V. (1998) Modeling of transport risk for hazardous materials. *Operations Research Chronicle*, **46** (5), 626–642.

Ferrey, S. (2010) *Environmental Law: Examples & Explanations*, 5th edn, Aspen Publishers Online, New York, NY.

Frynas, J.G. (1998) Political instability and business: focus on Shell in Nigeria. *Third World Quarterly*, **19** (3), 457–478.

Jackson, K.T. (1987) *Crabgrass Frontier: the Suburbanization of the United States*, reprint, illustrated edn, Oxford University Press US, New York.

Kelly, J. (2006) *The Great Mortality: An Intimate History of the Black Death, the Most Devastating Plague of All Time*, reprint edn, HarperCollins, New York.

Leonelli, P., Bonvicini, S., and Spadoni, G. (1999) New detailed numerical procedures for calculating risk measures in hazardous materials transportation. *Journal of Loss*

Prevention in Process Industries, **12** (2), 87–96.

Levine, J. (2006) *Zoned Out: Regulation, Markets, and Choices in Transportation and Metropolitan Land Use*, Resources for the Future, Washington, DC.

McShane, C. and Tarr, J.A. (2007) *The Horse in the City: Living Machines in the Nineteenth Century*, illustrated ed., JHU Press, Baltimore.

Nembhard, D.A. and White, C.C. (1997) Applications of non-order-preserving path selection to hazmat routing. *Transportation Science*, **31** (3), 262–271.

Nozick, L.K., List, G.F., and Turnquist, M.A. (1997) Integrated routing and scheduling in hazardous materials transportation. *Transportation Science*, **31** (3), 200–215.

Pet-Armacost, J.J., Sepulveda, J., and Sakude, M. (1999) Monte Carlo sensitivity analysis of unknown parameters in hazardous materials transportation risk assessment. *Risk Analysis*, **19** (6), 1173–1184.

Purdy, G. (1993) Risk analysis of the transportation of dangerous goods by road and rail. *Journal of Hazardous Materials*, **33**, 229–259.

Rose, A.Z. (2009) A framework for analyzing the total economic impacts of terrorist attacks and natural disasters. *Journal of Homeland Security and Emergency Management*, **6** (1), 9.

Samuel, C., Keren, N., Shelley, M.C., and Freeman, S.A. (2009) Frequency analysis of hazardous material transportation incidents as a function of distance from origin to incident location. *Journal of Loss Prevention in the Process Industries*, **22** (6), 783–790.

Schweitzer, L. (2006) Environmental justice and hazmat transport: a spatial analysis in southern California. *Transportation Research Part D: Transport and Environment*, **11** (6), 408–421.

Sherali, H.D., Brizendine, L.D., Glickman, T.S., and Subramanian, S. (1997) Low probability-high consequence considerations in routing hazardous materials shipments. *Transportation Science*, **31** (3), 237–251.

Sulijoadikusumo, G.S. and Nozick, L.K. (1998) Multiobjective routing and scheduling of hazardous materials shipments. *Transportation Research Record*, **1613**, 96–104.

Verter, V. and Erkut, E. (1997) Incorporating insurance costs in hazardous materials routing models. *Transportation Science*, **31** (3), 227–236.

Part Four
Security of Hazmat Transports: International Policies and Practices

12
Security of Hazmat Transports in Italy

Paola Papa and Luca Zamparini

12.1
Introduction

The likely momentous consequences of a fraudulent event associated with dangerous goods transportation (that can cause significant harm to both humans and the environment) and the increased perceived terrorist risk have attracted growing attention towards hazmat-transportation security worldwide. In this respect, the study of the volumes of hazmat transport flows, the rigorous application of the international requirements related to hazmat transport security, and the monitoring of the occurrences of unlawful events involving the transportation of dangerous goods, can contribute to reduce the risk that such events take place.

This chapter analyzes the major security aspects linked to hazmat transportation in Italy. The size of this sector seems to be quite considerable in this country especially with respect to the quantity of dangerous goods carried by road in comparison to those of other European Union members. In Section 12.2, the relevance of dangerous goods transportation in Italy by road, rail and sea is examined and some statistical updates and trends related to hazmat transport are presented.

Section 12.3 proposes a brief overview of the Italian legislation and of the principles underlying the regulation of dangerous goods' transport. The process of enforcing hazmat-transportation security is internationally based on the revision and on the harmonization of regulatory frameworks and legal settings. In Italy, the legislation concerning hazmat transport security mostly comes from International or European laws, which are periodically modified in accordance with the international conventions and agreements and to the related evolving legal requirements.

Moreover, the issue of hazmat transport security also appears relevant if recent events related to the seizure of dangerous goods are considered. Some recent cases related to hazmat transport security in Italy are reviewed in Section 12.4. These episodes occur quite often and are mostly linked to illegal dangerous waste dumping. Such a phenomenon has been especially critical in previous decades

Security Aspects of Uni- and Multimodal Hazmat Transportation Systems, First Edition. Edited by Genserik L.L. Reniers, Luca Zamparini.

(between the 1980s and the 1990s) when it assumed a transnational dimension. In that period, many industrialized countries used to export, often surreptitiously, dangerous waste to developing nations (generally sub-Saharan African countries). However, the most significant seizures of dangerous goods in Italy that have taken place in the latest years do not regard toxic waste but a huge quantity of explosives discovered in a container at the transshipment port of Gioia Tauro in September 2010, and the detection of a highly radioactive container at the Voltri Terminal of the port of Genoa in July 2010. The last section of the chapter provides some concluding remarks.

12.2
Economic Significance of Hazmat Transport in Italy

The weight of hazmat transport in Italy is quite significant as compared to other European countries. With respect to hazmat transport by road in 2008, Italy was the third European country, with a share of 14% of the total volumes in the EU-27. Moreover, France, Germany, Italy, Poland, Spain and the United Kingdom cover 75% of the total road transport of dangerous goods in the EU-27 (Eurostat, 2009).

Before analyzing the statistical datasets related to hazmat transport in Italy, it has to be pointed out that available data often refer to different classifications of merchandise categories. They are the NST/R classification for transport statistics defined at European level, and the international ADR classification. In order to provide an harmonization of transport data, the European Commission decided to also adopt the international ADR classification for EU transport statistics in 1998 (EU Commission, 1998).

A cross-analysis of the two classifications highlights that only four groups of goods of the NST/R classification can be included among the "dangerous goods" defined by the ADR classification. They are: "crude oil", "petroleum products", "carbochemicals and tars", "chemical products, excluding carbochemicals or tars".

Lastly, as maritime transport regards, available statistics consider only two categories of dangerous goods: petroleum products and chemical products.

The following subsections will present a series of statistics related to hazmat transport by road, by rail, and by sea in Italy.

12.2.1
Hazmat Transport by Road

The evolution of hazmat transport by road between 2001 and 2007 shows a 5.6% decrease in the first half of the period (from almost 122 millions tons in 2001 to nearly 115 millions tons in 2004), and a 15.4% increase in the second half of the period (reaching almost 133 millions tons in 2007). The overall increase is due to the growth of the volumes of chemical products, including carbochemicals and tars (Table 12.1). It should be stressed, though, that petroleum products represent the largest transported hazmat category that accounts for half of the overall hazmat traffic.

Table 12.1 Road freight transport of dangerous goods (tons).

Commodity group	Year		
	2001	2004	2007
Crude oil	374726	162342	382701
Petroleum products	63102413	60417781	60967704
Carbochemicals, tars	17444645	15613121	24613573
Chemical products, excluding carbochemicals or tars	40860481	38711974	46679541
Total	121782265	114905218	132643519

Source: Istat.

Table 12.2 National and international road freight transport of dangerous goods. Year 2006. In million ton-km.

Dangerous goods	National transport	International transport	Total
1. Explosives	58	2	60
2. Gases, compressed, liquefied, dissolved und. pressure	1259	54	1313
3. Flammable liquids	6380	481	6861
4.1 Flammable solids	50	8	58
4.2 Substances liable to spontaneous combustion	62	na	62
4.3 Substance emitting flammable gases (with water)	na	na	na
5.1 Oxidizing substances	177	na	177
5.2 Organic peroxides	18	na	18
6.1 Toxic substances	162	114	276
6.2 Infectious substance materials	167	na	167
7. Radioactive substances	na	na	na
8. Corrosive substances	1356	178	1534
9. Miscellaneous dangerous substances	231	4	235
Total	9920	841	10761

Source: Eurostat.

Looking at the kind of goods transported according to the ADR classification (Table 12.2), the large majority of these are flammable liquids (63.8%), followed by consistent quotas of corrosive substances (14.3%) and compressed, liquefied, frozen liquefied and resoled under compression gases (12.2%).

The large majority of dangerous goods carried by road are related to the internal market (92.2%) with a low percentage of international transport (7.8%). Flammable liquids represent the most significant category both at national and at

Table 12.3 Road freight transport of dangerous goods divided by distance and merchandise group. Year 2007.

	Short distance (<50 km)		Medium–long distance (>50 km)		Total	
	Millions of tons	Ton-km (×1000)	Millions of tons	Ton-km (×1000)	Millions of tons	Ton-km (×1000)
Crude oil	–	–	382 701	70 450	382 701	70 450
Petroleum products	13 530 561	385 788	47 437 143	7 240 796	60 967 704	7 626 584
Carbochemicals, tars	10 551 061	262 442	14 062 512	2 131 846	24 613 573	2 394 288
Chemical products, excluding carbochemicals or tars	13 180 829	245 161	33 498 712	10 017 318	46 679 541	10 262 479

Source: Istat.

international level, while the share of toxic substances is more consistent for international transport (13.6% of the total) rather than for the internal one (1.6%). Similarly, the share of corrosive substances internationally carried (21.2%) is considerably higher in comparison with that of national transport (13.7%), while the opposite occurs for gases (12.7% of the total national transported volumes versus 6.4% of the total international transported volumes).

In 2007, the total amount of hazardous-materials transported by road was nearly 8.9% of the total goods carried by road. The largest merchandise category was represented by petroleum products (46%), followed by chemicals products (35.2%), carbochemicals and tars (18.6%), and a very small component of crude oil (0.3%). Looking at Table 12.3, it can be observed that the volume of petroleum products carried by road is much higher for the medium–long distance (49.7%) compared to the short one (36.3%), while the carbochemicals and tars travel preferably over the short (28.3%) rather than over the medium–long distances (14.7%) (Table 12.3).

12.2.2
Hazmat Transport by Rail

Hazmat transport by rail accounted for an amount of 3.51 million tons of goods carried in 2007. The figures in Table 12.4 show an overall increase of 33.2% in the volumes between 2005 and 2007, mainly due to petroleum products (+55%) but also to chemical products (+15.2%). In 2007, the most common commodity group carried by rail was represented by petroleum products (56.3%), followed by a consistent share of chemical products (40%) and by quite a small quota of carbochemicals and tars (3.7%), while crude oil is not carried at all by rail.

The evolution of rail hazmat transport by merchandise category between 2004 and 2009 (Table 12.5) shows an increase in the volumes of flammable liquids, the largest transported commodity, since 2005. After a decrease in the first year (−8.3%), there has been a continuous positive trend that has reached a value of

Table 12.4 Rail freight transport of dangerous goods. Years 2005–2007.

Commodity group	2005		2006		2007	
	Tons	Ton-km (×1000)	Tons	Ton-km (×1000)	Tons	Ton-km (×1000)
Crude oil	–	–	–	–	–	–
Petroleum products	1 276 622	483 358	1 688 917	609 415	1 979 076	701 326
Carbochemicals, tars	141 498	41 834	135 847	40 986	128 558	37 866
Chemical products, excluding carbochemicals or tars	1 219 230	478 204	1 382 114	555 695	1 404 572	563 160
Total	2 637 350	1 003 396	3 206 878	1 206 096	3 512 206	1 302 352

Source: Istat.

Table 12.5 Rail freight transport of dangerous goods divided by kind of goods (Tons).

	2004	2005	2006	2007	2008	2009	'09/'04 Δ%
1. Explosives	10 296	2603	3315	4703	1534	5706	–44.6%
2. Gases, compressed, liquefied, dissolved under pressure	1 406 133	1 393 745	1 230 483	1 294 003	1 667 278	1 645 236	17.0%
3. Flammable liquids	2 318 011	1 431 198	1 626 028	1 821 464	2 082 409	2 984 132	28.7%
4.1 Flammable solids	131 198	111 630	98 795	121 092	113 508	172 520	31.5%
4.2 Substances liable to spontaneous combustion	313 825	45 906	14 964	12 519	12 020	8032	–97.4%
4.3 Substance emitting flammable gases (with water)	16 631	15 700	21 707	35 770	37 326	49 074	195.1%
5.1 Oxidizing substances	51 822	40 014	51 440	58 263	58 904	62 344	20.3%
5.2 Organic peroxides	2544	4280	3628	2294	3027	2855	12.2%
6.1 Toxic substances	470 991	482 242	520 416	454 065	362 690	485 415	3.1%
6.2 Infectious substance materials	29	–	–	–	–	–	nm
7. Radioactive substances	892	221	106	367	3893	1992	123.3%
8. Corrosive substances	675 963	682 050	633 034	525 616	524 205	675 736	0.0%
9. Miscellaneous dangerous substances	405 136	351 229	379 311	374 673	354 523	658 921	62.6%
Total	5 803 471	4 560 818	4 583 227	4 704 829	5 221 317	6 751 963	16.3%

Source: Istat.

2.98 million tons in 2009. The volumes of gases register a more stable trend, around 1.5 million tons for all the period. Similarly, corrosive substances reach nearly the level of 0.68 million tons both in 2004 and in 2009, with lower values in 2007 and in 2008.

The category that shows the most significant increase are substances emitting flammable gases with water (rising from about 17 000 tons in 2004 to about 49 000 tons in 2009), radioactive substances (rising from about 900 tons in 2004 to about 2000 tons in 2009) and miscellaneous dangerous substances (rising from about 405 000 tons in 2004 to about 659 000 tons in 2009). On the contrary, the most noteworthy decreases are those of substances liable to spontaneous combustion (falling from about 314 000 tons in 2004 to about 8000 tons in 2009) and of explosives (falling from about 10 000 tons in 2004 to about 6000 tons in 2009).

12.2.3
Hazmat Transport by Sea

The share of petroleum products in the international and short sea shipping is considerably higher than that of chemicals, both for the loaded (85.1% of the total in 2007) and for the unloaded cargo (92.4% of the total in 2007) (Table 12.6).

Looking at the volumes of international and short sea shipping between 1990 and 2007, it can be noticed that the unloaded cargo tonnage of petroleum products has followed a stable trend, ranging around 170 millions tons in all the period considered (with the sole exception of the higher amount of 177.424 millions tons registered in 1990). On the contrary, the volumes of loaded petroleum products are more unstable, registering an overall increase of 16.3% between 1990 and 2007, but low values in 1995 and 2000 (respectively, 44.159 millions tons and 45.304 millions tons), and a noticeable growth between 2000 and 2005 (+28.5%). Since then, the amounts remain mostly unchanged till the end of the period.

Unloaded chemical products rise continuously until 2005 and then remain stable around the level of 14 millions tons, while loaded chemical products register a considerable increase from 2000 and 2005 (11.5%), a slight rise in the last three years (2.9%) and an overall growth of 52.5% between 1990 and 2007.

Table 12.6 International and short sea shipping of petroleum and chemical products, by unloaded and loaded cargo (10^3 tons).

	1990	1995	2000	2005	2006	2007
Cargo tonnage unloaded						
Petroleum products	177 424	168 360	170 245	168 939	170 440	170 683
Chemical products	9399	9965	12 074	14 186	14 005	14 025
Cargo tonnage loaded						
Petroleum products	51 525	44 159	45 304	58 225	56 266	59 950
Chemical products	6870	5826	8385	10 187	9835	10 479

Source: Istat; Estimations by the Ministry of Infrastructure and Transport for 2007.

Table 12.7 Gross weight of seaborne petroleum products by country of origin or destination ranked according to 2008 data (Millions of tons – countries' ranking of previous years in brackets).

	2005	2006	2007	2008
Libya	36 049(1)	35 153(1)	38 522(1)	35 088
Egypt	29 902(2)	26 665(2)	24 173(2)	22 881
Russia	28 419(3)	25 473(3)	23 191(3)	22 446
Turkey	2326(14)	5071(7)	10 030(4)	16 103
Spain	4601(8)	4221(8)	5829(7)	6100
Ukraine	5687(6)	5826(6)	7818(5)	5776
Algeria	6107(5)	6712(5)	4602(8)	5576
Syria	5422(7)	4047(9)	5911(6)	3765
Malta	2052(15)	1817(17)	2073(14)	3200
Georgia	6551(4)	7523(4)	4114(9)	2838
France	3214(10)	2666(12)	2627(11)	2829
USA	2485(11)	2392(14)	3017(10)	2772
Nigeria	2475(12)	3767(10)	1275(19)	2284
Norway	1914(16)	2977(11)	2278(13)	2099
Greece	1038(19)	1158(22)	1182(20)	1784
Rest of the world	20 318	24 628	24 631	26 056
Total	158 560	160 096	161 273	161 597

Source: Authors' elaboration on Istat Database.

The main trade partners of Italy by origin and destination of seaborne petroleum products are Libya (21.7% of the total in 2008), Egypt (14.2% of the total in 2008) and Russia (13.9% of the total in 2008). They maintain their rank from 2005 to 2008 (Table 12.7). The seaborne trade of petroleum products is unsurprisingly quite concentrated for origins and destinations, with the four major trade partners accounting for a 59.7% share of the total amount of internationally transported petroleum goods.

The evolution of seaborne chemical products by country of origin and destination, on the contrary, highlights some changes in the major trading partners between 2005 and 2008 (Table 12.8). While Spain is stable at first place during the entire period, with the sole exception of 2007 when the main trade partner of Italy was the USA, in 2008 Turkey gains one position moving from third place of the previous years to second place, and France becomes the third trade partner in 2008, after have being in fourth place in 2006 and in fifth place in 2005 and 2007. The most significant rise between 2005 and 2008 is represented by China, ranking fifth at the end of the period for trade relevance. On the contrary, the most dramatic fall is that of Greece, passing from eighth place of 2005 to thirteenth place in 2008. However, it has to be considered that the differences in the volumes of seaborne chemical goods traded are quite small: in 2008, Spain, the first trade partner of Italy handled only the 8.3% of the total, and the joint importance of the four main trade partners weighted only 29% of the total.

Table 12.8 Gross weight of seaborne chemical products by country of origin or destination ranked according to 2008 data (Millions of tons – countries' ranking of previous years in brackets).

	2005	2006	2007	2008
Spain	1375(1)	1448(1)	1294(2)	1031
Turkey	1010(3)	1051(3)	1106(3)	917
France	845(5)	968(4)	859(5)	854
USA	1358(2)	1360(2)	1322(1)	787
China	319(15)	555(8)	806(6)	751
Brazil	780(6)	804(5)	865(4)	528
Slovenia	371(12)	454(13)	467(10)	456
Egypt	734(7)	528(11)	381(14)	441
Tunisia	348(13)	362(14)	355(15)	377
Israel	415(10)	289(17)	398(12)	349
Saudi Arabia	500(9)	530(10)	568(8)	349
Croatia	245(18)	300(16)	332(17)	340
Greece	619(8)	559(6)	470(9)	339
Singapore	178(21)	280(18)	306(19)	310
Hong Kong	332(14)	270(19)	282(20)	294
Rest of the world	5375	5505	6231	4248
Total	14 804	15 263	16 042	12 371

Source: Authors' elaboration on Istat Database.

12.3 The Italian Legal Framework on Hazmat Transport Security

Italian laws on the transport of dangerous goods mostly come from European or International laws, which are pre-eminent in the Italian legal system. Italian legislation is normally drawn in the cases where higher-order regulations have a general character and/or need to be specified. At the international level, the benchmark in the regulation of hazmat transport is represented by the UN Recommendations on the Transport of Dangerous Goods. Following these recommendations, which are periodically updated, international institutions competent for different transport modes have derived their own regulations that are normally mirrored by the Italian legislation. Thus, according to the transport mode and the related competent international institution, main hazmat transport regulation can be classified as shown in Table 12.9.

There are two international organizations regulating the international transport of dangerous goods by air: the International Civil Aviation Organization (ICAO) and the International Air Transport Association (IATA). The ICAO regulation is based on two pillars: the Annex 18 to the Convention on International Civil Aviation, containing the basic provisions, and the Technical Instructions, providing all the detailed instructions for the safe international transport of dangerous goods by air. The IATA Dangerous Goods Regulations can be considered as an operating

Table 12.9 International institutions and regulations related to hazmat transport security.

Transport mode	International institution	Regulations
Air transport	ICAO (International Civil Aviation Organization)	Annex 18–The Safe Transport of Dangerous Goods by Air–to the Convention on International Civil Aviation (or Chicago Convention)
		Technical Instructions for the Safe Transport of Dangerous Goods by Air (Doc 9284)
	IATA (International Air Transport Association)	Dangerous Goods Regulations
Inland waterway transport	UNECE (United Nations Economic Commission for Europe)	European Agreement concerning the International Carriage of Dangerous Goods by Inland Waterways (ADN)
	CCNR (Central Commission for the Navigation on the Rhine)	
Rail transport	OTIF (Intergovernmental Organization for International Carriage by Rail)	RID–*Règlement concernant le trasport International ferroviaire des merchandises Dangereuses* (Agreement related to international transport of dangerous goods by rail), part of the Convention concerning International Carriage by Rail (COTIF)
Road transport	UNECE (United Nations Economic Commission for Europe)	ADR–*Accord européen relatif au transport international des marchandises Dangereuses par Route* (European agreement related to the international transport of dangerous goods by road)
Maritime transport	IMO (International Maritime Organization)	IMDG (International Maritime Dangerous Goods) code, part of the International Convention for the Safety of Life at Sea (SOLAS) for transportation on the high seas, and of the International Convention for the Prevention of Marine Pollution from Ships (MARPOL)

manual of the ICAO Technical Instructions. All the internal legislation for hazmat transport by air is then adopted in Italy through decrees by the Ministry of Infrastructure and Transport.

With respect to road transport, the ADR Regulation constitutes the most important point of reference. It is updated every two years by the Working Party on the Transport of Dangerous Goods (WP.15).The present version refers to the ADR 2011, applicable as from 1 January 2011. The original ADR Agreement signed in Geneva on 30 September 1957 under the aegis of the UNECE, was ratified in Italy by the law No. 1839 of 12/08/1962. At the European level, the Council Directive 94/55/EC extended the application of ADR regulation from international to national transportation. This Directive was adopted in Italy through the Ministerial Decree of 04/09/1996 and, in the following years, any further modification in the EC legislation, mainly due to the need for adapting it to technical progress, was each time transferred to the Italian legislation.

Following the same scheme adopted in the case of the regulation of hazmat transport by road, hazmat transport by rail is based on the RID Regulation, revised every two years by the RID Committee of Experts. The RID regulation was ratified in Italy by the law No. 976 of 18/12/1984. Similarly to ADR, it initially applied only to international transport of dangerous goods by rail, but it was later extended to national transportation, through the EC Directive 96/49/CE. This Directive was adopted in Italy through the legislative decree No. 41 of 13/01/1999 and, as for road transport, any further evolution in the EC legislation, has always been acknowledged by the Italian legislation.

As for the other transport modes, the adoption of the IMDG Code implies that Italy has to ratify the international regulation on hazmat transport by sea through decrees of the Ministry of Infrastructure and Transport. The European Commission intervenes with appropriate directives with the aim of implementing the international standards and requirements throughout its area. Council Directive 93/75/EEC of 13th September 1993 concerning minimum requirements for vessels bound for or leaving Community ports and carrying dangerous or polluting goods was adopted in Italy through the Decree of the President of the Italian Republic No. 268 of 19/05/1997. The Directive established that the carriage of these goods had to be accompanied by a declaration of compliance with the international requirements. It determined the information that the operators have to notify to the competent authority of the Member State before departure of a vessel leaving a port in a Member State or on departure from the loading port as a condition for the entry into the port of destination or anchorage. It lastly set up the appropriate procedures to be followed in the case of an accident. Directive 93/75/EEC was later repealed and replaced by Directive 2002/59/EC, which has been amended by Directive 2009/17/EC establishing a Community vessel traffic monitoring and information system.

As regards inland transport, the European Agreement concerning the International Carriage of Dangerous Goods by Inland Waterways (ADN) does not apply in Italy since it refers to the transport across national borders that is not possible in the case of Italy. However, through the Italian legislative decree No. 35 of 27/01/2010, the Agreement has been extended also to national transportation of hazardous materials by inland waterways.

The recent Italian legislative decree No. 35 of 27/01/2010 has adopted the new indications and requirements related to the transport of dangerous goods by road, rail or inland waterway coming from the EC Directive 2008/68/EC.

Transportation of radioactive material is regulated by another international organization, the International Atomic Energy Agency (IAEA), which is involved in the promotion of high-level safety and security worldwide, through intergovernmental legal instruments such as conventions, codes of conduct, treaties and agreements. The IAEA's activities related to the transportation of nuclear and radioactive materials are oriented to enhance the safety standards and the prevention of radioactive waste as well as the security requirements against transport sabotage, developing a coherent international framework. In Italy, the transportation of nuclear and radioactive materials is regulated by the law No. 1860 of 31/12/1962 and by the legislative decree No. 230 of 17/03/1995 together with several other decrees defining the technical specifics, for each transport mode, that are necessary in order to be authorized to handle a shipment of these goods. According to the law No. 1860 (Article 5) and to the decree No. 230 (Article 21), transportation of radioactive materials has to be provided only by vectors authorized by decree of the Ministry of Productive Activities in accordance with the Ministry of Infrastructure and Transport, the Ministry of Interior and the ISPRA, (the former APAT – *Istituto Superiore per la Protezione e la Ricerca Ambientale* – Institute for Environmental Protection and Research).

Alongside the above-mentioned long-lasting European treaties regulating the hazmat transport by air, road, rail and sea, which follow the UN Model Regulation, two other European statutes need to be mentioned. They are: Directive No. 95 of 2002 "on the restriction of the use of certain hazardous substances in electrical and electronic equipment", adopted in Italy through the Legislative Decree No.151 of 25/07/2005 and the Regulation on chemicals and their safe use (EC 1907/2006) – REACH (Registration, Evaluation, Authorization and Restriction of Chemical substances).

12.4 Recent Italian Case Studies Related to Hazmat Transport Security

The most relevant hazmat-transport security-related events in Italy concern the illegal traffic of dangerous materials, including toxic waste. Such dangerous materials are transported for illicit purposes, accompanied by counterfeited documentation and without taking into account the necessary transport safety requirements.

In order to prevent and inhibit such illegal transportation of dangerous goods, the Italian *Guardia di Finanza* (a branch of the Italian police connected to the Ministry of finance and specialized in financial crimes and in illegal and dangerous materials smuggling into the country), operates in close cooperation with the custom authorities in order to perform random and aimed checks at all ports, airports and border crossings. Moreover, a dynamic surveillance system that extends the network of checks across the country, is also implemented.

Recent cases related to illegal hazmat transportation consist in seizures of illicitly loaded dangerous goods, hidden in containers or trucks. Many cases can be defined "environmental crimes" and regard toxic waste, including radioactive material. The problem of illegal traffic and waste dumping is quite serious in Italy. Illegal dumping includes waste materials that have been dumped, tipped or deposited onto private or public land against the law. When such a phenomenon develops on a large scale, big amounts of waste materials are dumped in isolated areas or even exported to be dumped in other countries.

The case of Somalia is well known: since 1991, when its government collapsed, the country has been at the mercy of the illicit nuclear waste trafficking made by European and Asian firms. The hazardous waste ranged from uranium radioactive waste to industrial, hospital, chemical leather treatment and other toxic waste. For nearly 20 years, these foreign firms dumped freely their nuclear and toxic wastes in Somali waters, or even on the beaches, without taking account of the devastating impacts on the health of the local population and on the environment.

In April 2010, the phenomenon of waste dumping in Somalia was brought up during an unclear episode regarding the Italian-flagged tugboat, the Buccaneer. The tugboat had been hijacked in the Gulf of Aden off Somalia's northern coast. However, initially, the Somali authorities declared that the Italian tug had been seized by the security forces since it was loaded with toxic waste material. The governor of the Sanag area claimed that the seizure of the ship took place to avoid the spilling of such dangerous waste in Somali waters. On the contrary, the ship-owner of the Buccaneer, entirely denied the allegation. After four months of negotiations among the Italian government, the Somali government and the local authorities, the crew – consisting of ten Italians, five Romanians, and one Croat – was finally released, but the episode still remains unclear.

The Italian crime organizations have played a pre-eminent role in this business, dealing in radioactive nuclear waste and even in the clandestine production of plutonium. Following a recent investigation by the Italian authorities, eight former employees of the Italian National Agency for New Technologies, Energy and the Environment (ENEA) are suspected to have paid the clan to get rid of 600 drums of toxic and radioactive waste from Italy, Switzerland, France, Germany, and the US. Such toxic and radioactive material has been dumped, between the 1980s and the 1990s, at unauthorized, nonsecure sites in southern Italy, Somalia and in the Mediterranean Sea. At present, the persistence of illegal waste dumping in Italy is witnessed by the occurrence of the seizures of toxic and radioactive waste.

Another important episode that has to be mentioned is the "ecological disaster" of the Lambro River in February 2010. After a sabotage at a former refinery, tons of oil were spilled into the Lambro, a river near Milan. The spill has seriously affected local vegetation and fauna, causing an ecological disaster without precedent for the Lambro ecosystem, an area very rich for the variety of birds and other wildlife.

12.4.1
Detailed Description of the Most Recent Hazmat Transport Security-Related Events

Recent episodes related to illegal hazmat transportation are summarized in Table 12.10 that takes into account all the most important cases that have happened between 2009 and the beginning of 2011. It should be stressed that the large majority of these events have taken place within the premises or in the proximity of some of the most important Italian ports. As has been emphasized in the pre-

Table 12.10 Recent events related to hazmat transport security in Italy (2009–2011).

Date	Place	Goods	Transport mode
February 2009	La Spezia	Toxic waste	Maritime transport
August 2009	Naples	Toxic waste	Road transport
December 2009	Taranto	Ammunitions	Maritime transport
January 2010	Ancona	Butane Gas	Road transport
February 2010	Trieste	Radioactive pellets	Road transport
February 2010	Palermo	Corrosive materials	Road transport
July 2010	La Spezia	Toxic waste	Maritime transport
July 2010	Genoa	Radioactive material	Maritime transport
September 2010	Naples	Toxic waste	Maritime transport
September 2010	Gioia Tauro	Explosives	Maritime transport
January 2010	Viggianello (Potenza)	Toxic waste	Road transport

ceding subsection, the illegal trade of toxic waste, both at the internal and international scale, constitutes an important share of hazmat transport security instances. This is due to the relevant economic savings that firms can obtain by using illegal and/or criminal logistic channels.

Among the latest events related to hazmat transport security, the most significant one is undoubtedly the seizure of seven tons of RDX explosives called T-4, a powerful military explosive, hidden inside a cargo container at the Italian port of Gioia Tauro, during an international operation led by the national intelligence, in cooperation with the antifraud central office of the Custom Authority. The container, originated in Iran, was transshipped at the port of Gioia Tauro bound for Syria and it was supposed to contain powdered milk, according to the transport documents. It was on board the cargo ship "Finland" of the Swiss–Italian MSC shipping company. Given that the port of Gioia Tauro lies in the territory controlled by the "'ndrangheta", the powerful crime organization of the Calabria region, the authorities initially believed that the explosive could be connected to local mobsters' criminal activities. However, the large amount of explosive seized persuaded regional police officials that the recipients would most probably be a large international terrorist organization. It was thus ascertained that the explosives were to be acquired by a Middle-East terrorist group.

Another episode connected to the transshipment port of Gioia Tauro is the discovery of a highly radioactive container at the Voltri Terminal of the port of Genoa. This container had originated from Sun Metal Casting in Adjman in the United Arab Emirates and was supposed to carry 18 tons of copper for an Italian customer. It was exported through the Red Sea port of Jeddah and transshipped via Gioia Tauro to Genoa. At present (March 2011) the situation is still posing a threat. The container has been isolated by using other containers filled with stones and water, but the authorities have not yet decided whether to open it by remote-controlled robot, or to remove it from the port by barge. They fear that the container could be a terrorist weapon and that its opening could activate an explosive device.

Two other episodes have taken place in the port of La Spezia. In the first one, the custom officers, jointly with the Finance Police, seized 28.33 tons of toxic waste. The toxic waste was discovered in a cargo container after the scanning operations. The container, originated from an Italian firm, was to be embarked for Tunisia. In the second case, the custom officers seized 67 tons of toxic waste, containing polluting oils. The load was bounded for Tunisia and Morocco.

Further episodes have been related to the ports of Taranto, of Palermo, and of Naples. In the first case, the custom officers, in cooperation with the Finance Police, seized a large set of weapons and ammunitions, hidden in a cargo container originated in Shanghai (China), that was supposed to be transshipped in Taranto and, then, sent to Greece. The commercial value of the illegal load was nearly 135 000 euros. In the port of Palermo, the custom officers and the Finance Police stopped a truck loaded with nearly 4 tons of highly corrosive materials. The load was carried on an inadequate vehicle and without the required authorization. The driver declared the truck contained packaged cargo. In the last case, the custom officers of Naples, together with the Finance Police of the same city, seized at the port of Naples 11 containers with 300 tons of toxic waste, which were to be embarked for Malaysia. In the proximity of the same port, the Italian police stopped a truck containing toxic waste, including a large amount of iodine 131, a highly toxic and radioactive material, generally a hospital waste. This dangerous material was to be illegally dumped in the province of Naples.

Other episodes have involved road transport. The custom officers of Ancona, in cooperation with the Finance Police, seized 2080 boxes containing 74 880 small tanks of butane gas. The goods, carried by a truck coming from Greece, were falsely described as "metals". Moreover, the custom officers of Trieste, in cooperation with the Finance Police of Udine, seized 23 tons of Ukrainian radioactive pellets coming from Hungary. The goods, which were to be imported in Italy, presented a radiometric anomaly 2.5 times higher than the normal background values. Lastly, the forest police of Viggianello (Potenza) stopped and seized a truck carrying toxic waste in the Pollino National Park, a national park located between Basilicata and Calabria, in southern Italy. The truck had the authorization only for carrying normal waste, but there was also toxic waste hidden among the normal one.

12.5 Concluding Remarks

The analysis of the statistical dataset and the review of the recent national cases associated to hazmat-transportation security confirm that dangerous goods transportation in Italy is an important issue.

In terms of volumes of cargo transported, petroleum products represent the most significant category for all transport modes, followed by chemicals. The proposed review of relevant fraudulent national events highlights, nevertheless, a predominance of seizures of toxic waste for both road and maritime transport. The problem of dangerous waste dumping is, indeed, still serious and is often linked to the activities of criminal organizations but, at the same time, the recent seizures of two containers full of explosives and of radioactive materials, respectively discovered at the port of Gioia Tauro and at the port of Genoa, shift the attention towards other dangerous substances and reinforce the necessity of ensuring an adequate level of control over all kinds of hazmat transport activities. The recent upgrading of security standards in terms of container scanning procedures and customs checks, together with the random controls performed by the Finance Police, are demonstrating their effectiveness in containing such illegal activities and increasing the overall level of hazmat transport security.

At the same time, the high degree of harmonization of hazmat transport security regulation at the international level, deriving from the adoption of the international treaties and agreements concerning hazmat transportation for every transport modes, is a noteworthy starting point in order to guarantee a reasonable level of security standards along the entire supply chain.

Bibliography

Decreto Legislativo n. 230 del 17 marzo 1995, Attuazione delle direttive EURATOM 80/836, 84/467, 84/466, 89/618, 90/641 e 92/3 in materia di radiazioni ionizzanti.

Decreto Ministeriale del 04/09/1996 Attuazione della direttiva 94/55/CE del Consiglio concernente il ravvicinamento delle legislazioni degli Stati membri relative al trasporto di merci pericolose su strada.

Decreto del Presidente della Repubblica n° 268 del 19/05/1997, Regolamento di attuazione della direttiva 93/75/CEE concernente le condizioni minime necessarie per le navi dirette a porti marittimi della Comunità o che ne escono e che trasportano merci pericolose o inquinanti, nonché' della direttiva 96/39/CE che modifica la predetta direttiva.

Decreto Legislativo del Governo n. 41 del 13/01/1999 Attuazione delle direttive 96/49/CE e 96/87/CE relative al trasporto di merci pericolose per ferrovia.

Decreto Legislativo del Governo n. 151 del 25/07/2005 Attuazione delle direttive 2002/95/CE, 2002/96/CE e 2003/108/CE, relative alla riduzione dell'uso di sostanze pericolose nelle apparecchiature elettriche ed elettroniche, nonché allo smaltimento dei rifiuti.

Decreto Legislativo del 27 gennaio 2010, n. 35 Attuazione della direttiva 2008/68/CE, relativa al trasporto interno di merci pericolose.

EU Commission (1993) Council Directive 93/75/EEC of 13 September 1993 concerning minimum requirements for vessels bound for or leaving Community ports and carrying dangerous or polluting goods.

EU Commission (1994) Council Directive 94/55/EC of 21 November 1994 on the approximation of the laws of the Member States with regard to the transport of dangerous goods by road.

EU Commission (1996) Council Directive 96/49/EC of 23 July 1996 on the approximation of the laws of the Member States with regard to the transport of dangerous goods by rail.

EU Commission (1998) Council regulation No 1172/98 of 25 May 1998 on statistical returns in respect of the carriage of goods by road.

EU Commission (2002) Directive 2002/95/EC of the European Parliament and of the Council of 27 January 2003 on the restriction of the use of certain hazardous substances in electrical and electronic equipment.

EU Commission (2006) Regulation (EC) No 1907/2006 of the European Parliament and of the Council of 18 December 2006 concerning the Registration, Evaluation, Authorisation and Restriction of Chemicals (REACH), establishing a European Chemicals Agency, amending Directive 1999/45/EC and repealing Council Regulation (EEC) No 793/93 and Commission Regulation (EC) No 1488/94 as well as Council Directive 76/769/EEC and Commission Directives 91/155/EEC, 93/67/EEC, 93/105/EC and 2000/21/EC.

Eurostat (2009) Global economic crisis hits European road freight transport in the fourth quarter of 2008, Statistics in focus, No. 86.

IATA (2011) Dangerous Goods Regulations, 52nd edn.

ICAO (1944) Convention on International Civil Aviation, Signed at Chicago, 7 December 1944.

IMO (2010) International Maritime Dangerous Goods – IMDG- code.

Istat (2005) Statistiche dei trasporti. Anni 2002–3, Istat, Roma.

Istat (2007) Statistiche dei trasporti. Anno 2004, Istat, Roma.

Istat (2010) Il trasporto merci su strada. Anni 2006–7, Istat, Roma.

Istat (2011) Trasporto ferroviario. Anni 2004–2009, Istat, Roma.

Legge n. 1839 del 12 agosto 1962 Ratifica ed esecuzione dell'Accordo europeo relativo al trasporto internazionale di merci pericolose su strada, con annessi Protocollo ed Allegati, adottato a Ginevra il 30 settembre 1957.

Legge n. 1860 del 31 dicembre 1962 Impiego pacifico dell'energia nucleare.

Ministero delle Infrastrutture e dei Trasporti (2009) Conto Nazionale delle Infrastrutture e dei Trasporti. Anni 2007–2008.

Ministero delle Infrastrutture e dei Trasporti (2010) Conto Nazionale delle Infrastrutture e dei Trasporti. Anni 2008–2009.

OTIF (2011) RID (Règlement concernant le trasport International ferroviaire des merchandises Dangereuses).

UNECE (2009) UN Recommendations on the Transport of Dangerous Goods – Model Regulations. Sixteenth revised edition, United Nations.

UNECE (2011) ADR – Accord européen relatif au transport international des marchandises Dangereuses par Route.

UNECE Committee on Inland Transport (2010) AND – European Agreement Concerning the International Carriage of Dangerous Goods by Inland Waterways.

13
Security of Hazmat Transports in The Netherlands from a Security Practitioner's Point of View

Henk Neddermeijer

13.1
Introduction

This chapter discusses some security issues of hazmat transports in The Netherlands. One can look at the subject matter from a myriad of perspectives, for instance from a political point of view, an economical or even from the viewpoint of an advisor on (the transport of) hazardous material, all in their own right important. However, we choose explicitly to address these issues from the point of view of a security practitioner. In this chapter we attempt to examine to what extent, if any, security is an issue in carrying out hazmat transports in The Netherlands. For the better understanding of the way *security* as a notion is used in this chapter, we explain it as "controlling risks *deliberately* caused by all sort of opponents". Thus, the nature and the extent of risks depend on three interacting aspects: "interests", "threat" and "the capacity to withstand the threat" (see Figure 13.1 – referred to as the "risk triangle"). We will relate these three aspects to hazmat transports in The Netherlands. Furthermore, at the end we will attempt to present a tentative conclusion about the situation of security of hazmat transports in The Netherlands.

First, we will discuss the various notions of security. It is used in many different ways – *security* as a management discipline, *security* as an aspect of policy and *security* as a hygiene factor (Herzberg, 1975). In this chapter our starting point will be "security" as a management discipline, and even within this clear preference we will see that the notion of "security" as used in "security management" is multiexplicable.

Then it will be demonstrated that The Netherlands is a risk-prone country with a risk prone infrastructure. Despite this observation, the Government stimulates international transport and logistics entrepreneurs to establish their operations in the Netherlands. It underlines the importance of the sector in The Netherlands and, when related to the high amount of chemical industries in the country, it includes the importance of hazmat transport being a significant part of that

Security Aspects of Uni- and Multimodal Hazmat Transportation Systems, First Edition. Edited by Genserik L.L. Reniers, Luca Zamparini.

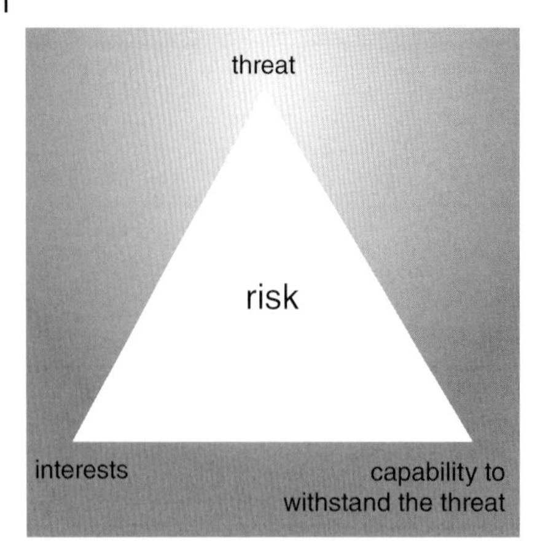

Figure 13.1 Risk triangle.

transport and logistics sector. Without considering the capacity to withstand the range of threats present at this very moment, it demonstrates that The Netherlands, being a risk-prone country with a risk-prone infrastructure, is probably more vulnerable than other member states of the European Union to safety and security incidents, accidents or disasters.

In this part we will make a detour to objective observations regarding the Dutch infrastructure, which is also a risk factor, and to the consequences of perceptions of transport and logistics risks (by definition subjective).

Subsequently, we will examine the transport and logistics in The Netherland in more detail, following by security issues regarding transport and logistics. At the macro level we will address the danger of terrorism, on the meso- and microlevel we will address a few aspects of crime in the transport and logistics sector. Next, the activity "safety first" will be discussed. This-often heard statement influences security in a positive way; however, is it enough to take safety measures to be secure? It will be demonstrated that many "stakeholders" are occupied with security–two relevant examples, the project "guarded parking areas" and the project "security lanes" will be discussed.

Finally, a reflection on security, security management and security measures in relation to transport in general, and hazmat transport in The Netherlands in particular, will be given as a careful conclusion.

13.2 Safety and Security

First, it is essential to define security in an unambiguous way, as far as this is possible. The reason for this is that over time, and in different places, the notion of security changes.

As mentioned in the introduction, we link *security* to risks *deliberately* caused by opponents. In order to manage these risks "security management" is used. This is a management area related to asset management, physical security and human-resource safety functions. It entails the identification of an organization's interests, such as assets and information, and the development, documentation and implementation of policies, standards, procedures and guidelines. Threat and risk analyses are pillars within the broad area of security management. Threat and risk are difficult concepts to define even for security professionals. Talbot and Jakeman (2009: xix) state that: "Confusion surrounding frequently interchanged terms such as threat and risk, likelihood, and probability is unlikely to go away, particularly as most of these terms not only are translated differently between languages but also reflect different cultural nuances". The question of definition is not unknown in the Dutch security field and includes the concepts of safety and security as well.

For example, from the practitioner's point of view, *security* can be explained as "the state of being free from danger or injury" or as others say, the "freedom from anxiety or fear". More concretely, *BusinessDictionary.com* defines *security* as "prevention of and protection against assault, damage, fire, fraud, invasion of privacy, theft, unlawful entry, and other such occurrences caused by deliberate action". This dictionary explains *safety* by stating that it is the "relative freedom from danger, risk, or threat of harm, injury, or loss to personnel and/or property, whether caused deliberately or by accident". In general, Dutch security practitioners disagree with the clause in this definition "whether it is caused deliberately or by accident". In their perspective, safety is all about human failure, human error, accidents and natural disasters, whereas security is about *intentional* actions taken to cause any kind of harm. One can, however, agree about the consequences resulting from an incident, accident or a disaster: for the general public it hardly matters what causes an incident, accident or a disaster. Figure 13.2 illustrates this viewpoint. To a certain extent they will probably suffer from such an event. In this

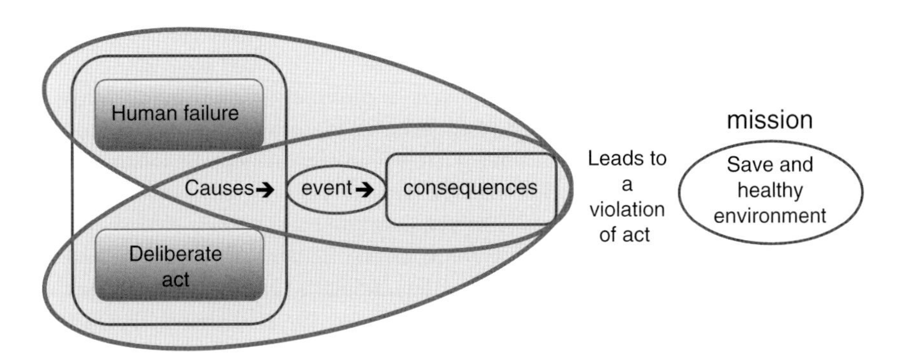

Figure 13.2 Modified bow tie.

chapter security will be approached from the viewpoint that an incident or a disaster deliberately can or will be caused. In this way, the threat of adversaries aimed at hazmat transports is an important issue to pay attention to.

13.3 The Netherlands: a Risk-Prone Country with a Risk-Prone Infrastructure

The Netherlands' vulnerability for safety and security incidents, accidents and disasters will be demonstrated by describing a flood and subsequently a fire in a chemical plant. Although this last incident was safety related, one can imagine that such an incident can be triggered deliberately. But first, let us examine The Netherlands as a country within Europe.

On Tuesday 8 February 2011, 19:37 GMT +1, the population of The Netherlands numbers 16 666 512 inhabitants. This number calculated by Statistics Netherlands (CBS – Centraal Bureau voor de Statistiek) for a given moment is partly based on an estimate. The fact is, however, that this number of inhabitants are living in a country that measures 41.526 km^2, and so The Netherlands has a high population density of about 400 inhabitants per square kilometer.

This density figure even has to be corrected because the Dutch territory consists of more than 18 per cent water. It is important to recognize that a large part of the country lies below sea level and most of its inhabitants live below that level. It is for this reason that the Dutch dikes are world famous. The story of Hansie Brinkers of Spaarndam is a worldwide well-known saga; the boy who saved his little village "Spaarndam" from a flood by plugging his finger into a dike. Jonkman, Vrijling, and Kok (2008, p. 1357) stated in "Flood Risk Assessment in The Netherlands: A Case Study for Dike Ring South Holland" that "Large parts of The Netherlands are below sea level. Therefore, it is important to have insight into the possible consequences and risks of flooding".

Flood was and still is a real potential disaster for The Netherlands. Since the floods of 1953, the Dutch government has put in place extensive defenses against the sea, and substantial legislation – resulting in the Dutch Disaster Act of 1985. This act places obligations upon organizations and institutions in this country that have an impact at the level of *security* practitioners. Nowadays and in general, *preincident prevention* (intelligence and protective security) and *postevent response* (emergency management and *in casu* business continuity management) are two concepts that resonate in the world of security. To plan for eventualities is to try to foresee them. The Dutch can foresee the eventuality of flood by looking over the nearest sea-wall. However, the "old eyes" are not very innovative and have been complemented by other more sophisticated instruments.

This consideration about the vulnerability of The Netherlands regarding *Acts of God*, such as a flood, is encouraged by safety thoughts. From a different – security – perspective one can say that the vulnerability of The Netherlands equals the vulnerability triggered by safety-related incidents.

The foremost of the inhabitants live in the Randstad or Randstad Holland. The Randstad is a conurbation where about 7 million people are living, working and relaxing. The cities of Amsterdam, Rotterdam, Utrecht and The Hague are part of this conurbation. The port of Rotterdam is the largest harbor in Europe; situated near Amsterdam Schiphol Airport. It goes without saying that most of the industry, including the chemical industry, can be found in the Randstad. An incident in such a crowded environment can easily evolve into a megadisaster. The second example, a fire in the chemical plant Chem Pack in Moerdijk, just below the city of Dordrecht in the Rotterdam Area, demonstrated recently the consequence of a such a calamity. At Chem Pack, flammable, poisonous and corrosive agents were stored and distributed. On 5 January, 2011 a fire started and destroyed the chemical plant completely. Employees of Chem Pack stated that the fire started by filling a tank truck. The fire caused a lot of explosions and the wind blew a gigantic column of black smoke in the direction of Dordrecht and Rotterdam. At some point, one prediction was that the smoke would move in the direction of Amsterdam and Ijmuiden, and of Utrecht as well (Timofeeff, 2011). If this prediction had become true, the whole Randstad area – 7 million people – would have been affected by this fire in Moerdijk, south of Rotterdam. Of course, people in the area Moerdijk and Dordrecht were advised to close doors and windows.

Though luckily in this incident there were no fatalities and casualties, the material damage was enormous. At the time of the fire the maritime transport on the "Hollands Diep" and a part of the highway A17 were closed down resulting in an massive traffic jam. Another traffic jam of eight kilometers was caused by "disaster tourists". Air transport, however, experienced no affect. In the aftermath of the fire the Mayor of the city of Breda, situated in the neighborhood of Moerdijk, estimated the total damage at € 40 to 60 million. He calculated this in his function of vice president of a related Safety Region, a partnership of local governments, fire departments and municipal health services.

One can easily imagine that a carefully positioned bomb somewhere in the Rotterdam Botlek area, where a lot of oil and (petro)chemical industries are situated, with a favorable wind in the right direction, can cause a high total of fatalities and casualties. When thinking about this awful scenario, one can imagine that security precautions have been made – adversaries will not easily be able to execute such a plan.

13.4
The Dutch Transport Infrastructure as Risk Factor

It is clear that the traffic problems due to the Chem Pack fire mentioned above have demonstrated the vulnerability of the Dutch transport infrastructure. An incident – small or large – can easily influence the business continuity of transport and logistics in the Netherlands and will – to some extent – affect the population. Does this objective observation resemble a subjective perspective on this matter?

In other words, does transport and logistics influence the risk perception of the population?

The Netherlands have long been faced with traffic jams. Moreover, the complexity of routes for transport modes (road, rail, aviation, shipping and pipelines) is high, caused by the large amount of transport infrastructure in a small area. This being not a new problem, a department of the Ministry of Infrastructure and the Environment, *in casu* the Kennisinstituut voor mobiliteitsbeleid – KiM (Knowledge Institute for Mobility Policy, translation HN), researches the importance of mobility and transport to the Dutch population in a so-called *knowledge line*. On the whole, the institute researches developments in society, the behavior of citizens and companies and the way in which mobility, economy, environment and *security* interact.

Regarding security issues, one can imagine that transport and logistics will interact with the risk perception of the population. Day by day, literally 24/7, the Randstad experiences those traffic jams. Most often one can speak of a traffic congestion. Public and private traffic are confronted with rows of trucks and lorries driving on the highways to the port of Rotterdam. Alongside the highways in de Randstad the exposure time of potential dangerous freights is long. Research of Visser and Hendrickx (2000, p. 46) demonstrated that this interaction between transport and logistics and risk perception exists, although citizens in general do not dwell on the risk of (intermodal) freight transport. In group discussions organized by these scholars it appeared that "participators found that freight transport is getting more unsafe, though still is safe and secure".

Comparing with the other transport modes – rail, aviation, shipping and pipelines, the respondents consider road transport as the most unsafe way of transport because of the perceived individual risks. Visser and Hendrickx found that "the type of goods that is transported plays an important role in risk perception" and that the participators think that "the more dangerous transport is, the better it is organized and secured".

One can easily understand this general notion of risk perception, because of the fact that the transport sector must comply with numerous national and international laws, bylaws and regulations. This legislation is an important contributor to the capacity to withstand threats in the transport and logistical sector. Of course, the national Dutch transport legislation meets the international standards. In particular, bearing these regulations in mind, and keeping in mind that the awareness of organizations and drivers involved in hazmat transport is strongly developed, the participators' assumption is significantly true.

Despite the discussed objective and subjective point of view on risks, however, one can perceive that risks of transport and logistics will continuously change, both in time and in place. This can, and will, result in the need to change legislations. However, in the public domain and in a group consisting of involved conveyors as well, it seems that there exists a kind of reservation concerning the development of new legislations and regulations, or the improvement of existing national legislations and regulations. That is to say, the government's and the industry's greatest concern is the fear of disturbing the fragile balance on a "level playing field" in Europe. Any disturbance of this balance by national actions to change legislation by introducing (additional) security measures in laws or something similar will not be accepted. The national focus is on what they do in Europe; "Waiting for Europe" is therefore a many heard respond.

From a security point of view one can conclude that the density of people in The Netherlands as a whole and the Randstad in particular, combined with the high extent of industrialization, among which the chemical industry, the number of logistical movements and the complexity and vulnerability of transport routes for all modes will determine The Netherlands as being risk prone and susceptible to deliberately caused incidents.

13.5
Transport and Logistics in The Netherlands

Still discussing the aspect of "interest" as one of the three contributing components of the "risk triangle" we have to research the significance of the value of transport and logistics for The Netherlands. Over centuries, The Netherlands has changed from an agricultural and commercial economy into an industrial society. However, the economic development did not end here. Nowadays, most of the inhabitants are working in the tertiary sector (also known as the service sector or the service industry). The tertiary sector can be regarded as the largest sector of the economy in the Western world, and it is also the fastest growing one. Transport and distribution are important segments within this service sector. From an economic point of view, one can conclude that the transport and logistics sector represents a huge interest for countries and entrepreneurs within those countries. Thus, national and international transport and logistics are of utmost importance to the economy of The Netherlands and therefore a fundamental theme in governmental policy. Furthermore, the government regards Logistic and Supply Chains as an important part of the national innovation program. This becomes noticeable in the way the Dutch government is trying to persuade foreign (transport and logistical) entrepreneurs to locate their headquarters in The Netherlands. The Holland International Distribution Council (NDL/HIDC), which represents the transport and logistics sector in The Netherlands, states on their website:

> "The Netherlands: Your logistics gateway to the European market . . . and beyond. The Netherlands is famous for its expertise in international trade, and excellence in transport and logistics. [...] Many international companies use the Netherlands as location for their European supply chain operations; as point of entry into the EU market for intercontinental freight by sea and air, or as location for central-, bulk- or regional distribution centres."

As a sort of marketing message they mention that "The NDL/HIDC helps international companies make a smooth entry into the European market through the region's leading gateway, The Netherlands". This is not a new development. From a historical perspective, which was mentioned in a 2009 research commissioned by the Department of Transport, the Holland International Distribution Council concluded:

> "The Netherlands, land of traders. That is more than a motto; it is in our genes. Since the sixteenth century Dutch explorers sailed the seas in search of new countries and people, products and markets. In the 'Golden' seventeenth century The Netherlands had trading posts in any continent: distribution centres that served as transhipment points between coast and hinterland, between The Netherlands and the world. Trade flourished by a combination of entrepreneurship, venture and good logistics management." (NDL/HIDC and TNO Business Unit Mobiliteit en Logistiek, 2009: 5) (translation: HN).

Since the seventeenth century, the importance of transport and logistics has been extended to almost unprecedented levels. The following facts will illustrate this point. In 2007 the Dutch share of the world trade was 3.7% and with that, The Netherlands occupied seventh spot in the world ranking, behind France and Great Britain, though before Italy, Belgium and Canada. In 2004, The Netherlands ranked eighth (NDL/HIDC and TNO Business Unit Mobiliteit en Logistiek, 2009: 9).

Besides this increase in growth, the character of the business in the Netherlands has changed. In the last 10 years the Dutch imports and exports show a more global character than before. The yield of these logistical activities is € 40 billion and 750 000 jobs in 2007 (NDL/HIDC and TNO Business Unit Mobiliteit en Logistiek, 2009: 15). When regarding the Dutch transport market as a part of the logistical sector, the volume of this market exceeded € 20 billion in 2007 (NDL/HIDC and TNO Business Unit Mobiliteit en Logistiek, 2009: 17).

At the opening of a new Dutch warehouse in October 2010 for the Japanese organization NRR Global Logistics Netherlands BV, the managing director Chris Coome confirmed this importance of the transport and logistics sector for The Netherlands. He stated that "The Netherlands is the only European country that can rightfully call itself 'the gateway to Europe' and a major hub to the rest of the world". In his opening speech he mentioned, that "there are other reasons for being pleased with having chosen The Netherlands". He stated that "from the beginning the support we received from the Dutch Embassy in London, the NFIA [Netherlands Foreign Investment Agency], as well as from organizations such as HIDC, SADC [Schiphol Area Development Company] and the Customs department, has accelerated and simplified the entire location process a great deal".

Not only organizations compliment The Netherlands on its transport and logistics policy, the Dutch population also have a comparable positive image of this sector, as was stated clearly in the research of Visser and Hendrickx (2000, p. 46).

Another tribute comes from the Dutch Union for the Chemical Industry. This "Vereniging van de Nederlandse Chemische Industrie" (VNCI) states on their website, that:

> "The preconditions in the Netherlands create a favorable investment climate for the chemical industry. Important raw materials are available or easy to supply, while an extensive transportation network provides access to the European market. Furthermore, chemical research and training in the Netherlands are among the best in the world.
>
> Together with the Dutch culture and mentality, they form a powerful chemical industry that boosts the economy and takes the lead in sustainable development and entrepreneurship." (www.vnci.nl)

The VNCI noticed that "about three quarters of the chemical products produced in the Netherlands are exported. Three quarters of that go to countries within the EU, and one quarter goes to non-EU countries. [...] In 2009, exports amounted to approximately € 60 billion. That is more than 19% of all of the exports from the Netherlands". One can conclude that, for this relatively small country, the transport and logistical sector is significant and that the transport and logistics of chemical products contribute essentially to this interest. It goes without saying that not all of the chemical transports are hazardous, but when looking for risks and even hazards one can easily find some in this sector. The *Knowledge Institute for Mobility Policy* (Kennisinstituut voor Mobiliteitsbeleid, translation HN) in close co-operation with the Ministry of Transport, Public Works and Water Management explained in their "Forecast regarding the transport of hazardous materials by road" in 2007, that most of the transports (in millions of tons) are carried out on the Dutch canals and rivers (spicks area in Figure 13.3), a small part on roads

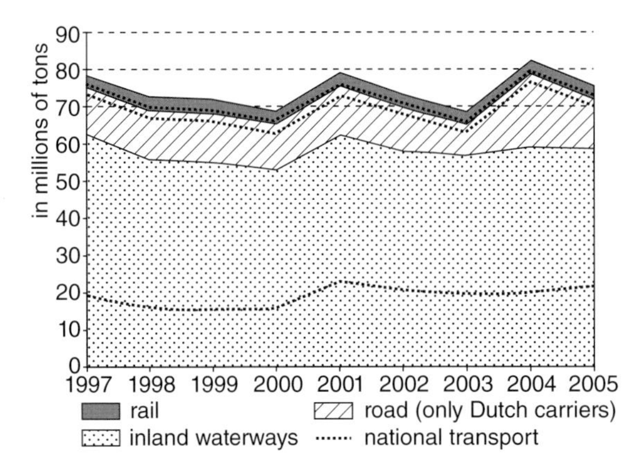

Figure 13.3 Transport of hazardous materials.

(stripes area in the figure) and even less by train (gray area). Reading this graphic, one has to keep in mind that more than 10 million tons of hazmat transport on roads in such a small country as The Netherlands is an enormous logistic quantity.

13.6 Security Issues in Transport and Logistics

Transport and logistics are of great interest to The Netherlands. The public private partnership *RPC/CrimiNee!* (translation; crime no!) recognizes and confirms this significance and value and states on their website:

> "If there is one thing that the Netherlands is good in, it is the transiting of goods. From the economic point of view, this sector has grown more rapidly than any other in the last twenty years. The presence of Schiphol airport, the ports of Rotterdam and Amsterdam and an excellent infrastructure, among other things, have enabled this impressive development. But this success is now under threat. Lorries are increasingly targeted by criminals who are after the valuable cargo. They usually strike at rest areas located along motorways."

The *RPC/CrimiNee!* brings regional partners together and constitutes the basis for projects and prevention oriented action plans aiming to reduce criminality where business is involved. Confirming the interest of the transport and logistics sector of The Netherlands, at the same time the *RPC/CrimiNee!* warns this sector for cargo theft as a major incident that can and will happen. The risk of cargo theft is one of many threats that the transport and logistical sector is exposed to.

The question to address now is in what way adversaries can disturb transport; in other words: what are the threats? This question addresses the second of the three contributing components of the "risk triangle".

In carrying out security management Dutch security practitioners often follow a number of well-known principles. The first of these is to question the "need to secure". Regarding the interest and the threat attached to The Netherlands being risk-prone nation and The Netherlands infrastructure as being risk prone as well, and – related to the presence of the existing threat for transport and logistics – the answer to this question is undoubtedly in the affirmative. Concerning a general motif, one can see that adversaries aspire to the following issues:

- theft of cargo;
- theft of a lorry/truck;
- theft of fuel;
- theft of a part of the lorry/truck;
- violence and threats;
- suspicious circumstances;
- other forms of transport criminality.

13.7 Terrorism

Literally looking at other transport modes, stealing a locomotive or a ship is not very obvious, stealing an airplane (for instance by highjacking) is more imaginable–although in The Netherlands the following saying is opportune: "never say never". Besides, one has to keep in mind that the intentions of adversaries play an important role as well. One will remember 9/11 where high jacked airplanes are used as bombs, without saying, an act of terrorism. It is likely thinkable that a highjacked hazmat transporter can be used as a "bomb". What do the branch organizations say to this idea? Most important is, however, what the governmental intelligence and security services state this possibility. Using means of transport by terrorists in The Netherlands as a way to attack the country is conceivable. In order to be able to alert different sectors of industry and the operational services in time in the event of a heightened threat of a terrorist attack, the Dutch National Coordinator for Counterterrorism (NCTb) has introduced "levels of alert" for the economic sectors. In this alert system the following 15 sectors participate: seaports, oil, chemical, drinking water, electricity, telecom, gas, nuclear, financial, railways, municipal and regional transport, public events, hotels, tunnels and flood defences and airports. The NCTb states that "the system is aimed solely at professionals who may have to deal with a terrorist threat and enables prior measures to be taken quickly in order to minimize the risk of terrorist attacks in the Netherlands and to limit the potential impact of terrorist acts". One can see that in this alert system all transport modes participate. Such a system increases the resistance against terrorism–in other words "it contributes to the capacity to withstand the threat", being the third of the three contributing components of the "risk triangle".

13.8 Transport and Logistics Crime

Changing the focus from terrorism to crime, the above-mentioned examples of cargo theft are a major problem in The Netherlands. The government, the transport and logistical sector, and the branch organizations are tackling this problem fiercely. By taking the right security measures to withstand cargo theft, to some extent these may contribute to withstand terrorists as well. And of course as a spin off, measures taken for the transport sector will affect the hazmat sector as well. "The security approach to the issues mentioned above is worthwhile to monitor" according to the advice of the "Project Group Monitoring" related to the "Covenant Approach to Crime in the Transport Sector" (2010: 3). Regarding deliberate disruption of transport and logistical processes, the incidence of cargo theft is high. The Dutch transport insurance company TVM, for instance, estimates that the 2011 damage of theft from trucks will increase to € 500 Million. It is noticeable and to a certain extent understandable that all parties involved with prevention of transport crime, in one way or another, take their responsibility. To comprehend this, it is necessary to know that since 1992 the government has stimulated one's own responsibility for security. Since then, numerous action plans have been executed, focusing on different aspects of crime and crime prevention. The slogan that was used was "partners in security". Nowadays, this principle of responsibility is followed by a significant proportion of the population and organizations.

The government states on their website "*Rijksoverheid*" that the legislation of transport (of hazardous agents) in The Netherlands is mainly based on international agreements. Not only does there exist a strong European focus, a global perspective is present as well. One can reason that this is a correct approach, since the logistic and supply chain is the focal point of transport and logistical policy. As we will see next, the consequence of this focus is an absolute and strong orientation on international law, bylaws and regulations.

When focusing on the transport of hazardous agents, it is comprehensible that the "Accord européen relatif au transport international des marchandises dangereuses par route" (ADR, transport mode: road), the "Accord européen relatif au transport international des marchandises dangereuses par voie de navigation" (transport mode: ship), the "Règlement concernant le transport international ferroviaire des marchandises dangereuses" (transport mode: train), and the "International Civil Aviation Organization Technical Instructions" play an important role in The Netherlands legislation. Maritime regulations are numerous: SOLAS (International convention for safety of Life at sea 1974), MARPOL (International Convention for the Prevention of Pollution from Ships), IMSBC-Code (Bulk Cargos Code), BCH-Code (Bulk Chemical Code), IBC-Code (International Bulk Chemical Code), Vervoer van gas in tankschepen: GC-Code (Gas Carrier Code), IGC-Code (International Gas Carrier Code) voor gastankers die gebouwd zijn na 1 juli 1986, INF-Code (Irradiated Nuclear Fuel Code), Code (International Maritime Dangerous Goods Code). All of this legislation has been in whole or to some extent included in Dutch legislation. In the framework of this chapter it is relevant to mention the Dutch law "Wet vervoer gevaarlijke stiffen" (the Law on Transport of Hazardous Goods).

13.9
Safety First

These law, bylaws and regulations have one common characteristic: the content is primarily focused on safety. It is well known that some safety measures ensure a safe transport, and as a spin-off contribute to a secure transport. For example, because of safety regulations a lot of freight train wagons are double skinned. This modification *casu quo* adaptation gives the fire department more time to react to a fire. From a security point of view, however, double skinned means protection against for instance RPGs.[1]

1) Rocket propelled grenade, for instance the RPG-7. The RPG-7 is adaptable to the needs of both pirates and terrorists.

One can assume that to a minor extent security aspects specifically are addressed in these regulations (see for instance Chapter 1.10 of the ADR where this occurs). However, more and more the awareness arises that international security legislation is necessary. There are though still few examples for this awareness. The International Ship and Port Facility Security Code amends the Safety of Life at Sea (SOLAS) Convention (1974/1988) on security arrangements for ships, ports and government agencies. Schemes like Authorized Economic Operators, developed by the customs, present benefits to participators in international transport – however, the scheme demands a minimum level of security to be taken. Despite this limitation, one can state that particular laws, still to a modest extent, laws, bylaws and regulations aiming security contribute to the “capacity to withstand the threat”.

As mentioned before, one will recognize that theft of a hazmat transport vehicle can have serious consequences. The thought of using such a vehicle as a trigger for an attack, for instance by shooting a RPG into it and thereby causing an explosion, is considerably frightening. The extent of the reality and feasibility of such an scenario, however, is questionable.

On this level the government is working on the Basic Network (Basisnet, Figure 13.4 shows the design for the network in the north of The Netherlands). This Basic Network will be used as transport routes for hazardous transport. Along these routes, security zones will be arranged and no or limited construction will be allowed. In 2012 this Basic Network will be part of the Dutch legislation. This clear

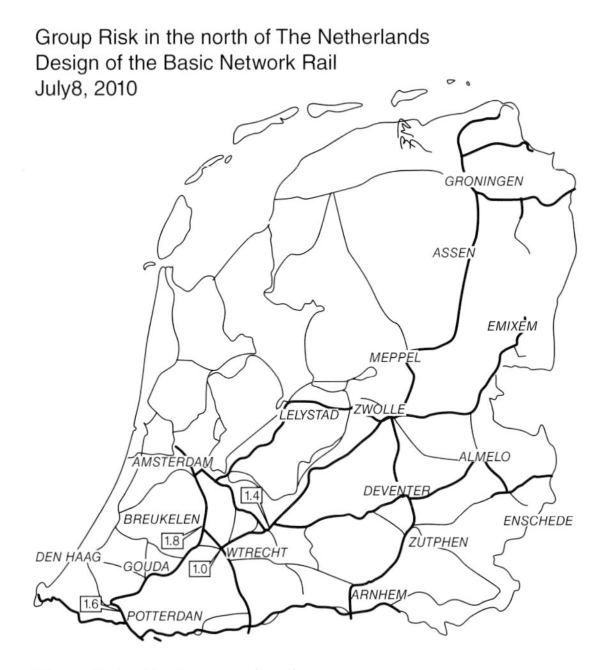

Figure 13.4 Basic network rail.

example of risk reduction shows us an approach related to "interest" as part of the aspects that contribute to "risk". Furthermore, above we spoke about cargo theft and stealing a truck as a whole. However, nothing was said about the drivers. In the government policy, attention is given to the security of drivers as well.

13.10
Partners in Security

Obviously, the Dutch government plays its role as a "partner in security". What other partners can we observe in the field? There are a lot of involved branch organizations and other organizations such as Air Cargo Netherlands, EVO, Koninklijke Nederlands Vervoer (Royal Dutch Transport, translation HN), Stichting Aanpak Voertuigcriminaliteit (Foundation Approach Vehicle Crime, translation HN), Transport en Logistiek Nederland (Transport and Logistics Nethetlands, translation HN), Transported Asset Protection Organization and the Verbond van Verzekeraars (Union of Insurers, translation HN). These partners joined in a Covenant Approach to Crime in the Transport Sector. The aim and focus of this covenant is to map transport criminality in a better way, and to develop countermeasures. The well being of drivers is a main concern in this covenant as well. This covenant has four important themes, (i) monitoring (awareness about where to take security measures and why), (ii) prevention (security starts by preventing incidents), (iii) repression (the [re]action of police and prosecutors on transport criminality), and (iv) international (measures taken in border areas).

One of the projects is to realize safe and secure parking lots and business premises, knowing that much cargo theft takes place on these premises. In line with this strive, The Ministry of Infrastructure and the Environment and its partners in the covenant have built a network of secured parking areas for transport along the main road network. Nowadays, there are about 25 parking areas in The Netherlands that are safe and secure to a certain level. These areas are labeled with a LABEL certificate (the level of security is indicated with padlocks – see Figure 13.5, the service level with stars).

Regarding this particular public private partnership, another thing is worth mentioning here. Despite the safe and secured parking areas mentioned above, many truck drivers are still resting along highways in resting areas. Therefore, the public private partnership CrimiNee! is developing a high-tech system that is called Secure Lane. The resting areas involved in this project are equipped with cameras. The number plate of every vehicle that drives to the resting areas is registered by cameras and surveillance cameras are monitoring the area for suspicious activities. The signals are real-time monitored in a regional monitoring center, and if something is wrong the images can be sent to the police and fire department.

The system has been operational for more than a year, and the incident rate dropped dramatically. In the end state there will be a security lane from the Belgium and German border to Rotterdam. There is, however, a reverse to every

Figure 13.5 Sign secured parking with padlock indication. Photo: Ministry of Infrastructure and the Environment.

medal. Sadly enough a displacement effect is manifest. Because of the successful approach to cargo theft on the route Venlo–Eindhoven–Rotterdam, the number of cargo thefts on the resting areas along the highway A2 and A76 in Limburg increased enormously in 2010. Last year 77 incidents were counted compared with 11 in 2009 (www.brabant.criminee.nl).

One can research the preventive activities of the conveyors themselves as well. Often, there are personal and financial circumstances that lead to not taking security measures by these firms. The margins on transport activities are small, the costs are high and in this situation a conveyor has to operate. An important instrument that can help such organizations is the ISO/PAS 28001 standard; "Security management systems for the supply chain–Best practices for implementing supply chain security–Assessments and plans". The branch organizations provide advice for their members as well. However, it is curious that, noticing multiple public private partnerships with comparable targets, some branch organization websites regarding the issue of security and security advice are blocked for nonmembers.

Often, an incentive coming from outside the firm will persuade people to do certain things–for instance to take security measures. A significant example is the insurance company TVM. In order to be able to control cargo theft, TVM has made particular conditions for the insurance of trucks and cargo. Part of these are security demands. The insurance company uses a risk-classification method in which a connection has been made between the type of cargo and security measures. For example, chemical products score on category III on a scale of I to IV.

Although chemical products do not entail the highest category, one has to bear in mind that this risk classification only focuses on cargo theft and not on other incidents. Of course, as we mentioned before, one can build a scenario of stealing a truck by terrorists loaded with hazardous materials. Again, the reality and the feasibility of that scenario is questionable.

In every case, the severity of cargo theft as an issue was and is high enough to appoint a national public prosecutor. This functionary has to coordinate the fight against transport criminality. With that, the importance of this issue has been established in The Netherlands.

13.11 Conclusion

Security in this chapter has been approached from the viewpoint that all proactive, preventive and reactive activities in this field are aiming to control the risks of deliberately caused incidents and disasters. Three contributing factors construct these risks; "interests", "threat" and "the capacity to withstand the threat" (referred to as the "risk triangle").

Regarding the factor "interests", transport and logistics are important in and for The Netherlands. A significant part of the national income depends on these sectors. One can notice this significance in the number of actors involved in this business; conveyers, branch organizations, insurance companies, governmental agencies and so on. Besides this, the government made logistic and supply chains important aspects of the national innovation program. Obviously, the interest of this business for The Netherlands is large.

The Netherlands as a small, very industrialized country, with a high population density, and complex infrastructure, is a vulnerable, risk-prone country. It is important to recognize, because the use of a tank truck filled with a hazardous agent as a bomb probably will have severe consequences for the population and the infrastructure. Of course, this threat – as a second contributing factor – will be constantly monitored by the Dutch intelligence and security services and the National Coordinator Counterterrorism. One can count this monitoring activity to the third contributing factor; the capacity to withstand the threat. Other activities are the enforcement of legislation (macrolevel), and the activities of the transport and logistical sector (mesolevel) and the entrepreneurs (microlevel).

Hazmat transports have to comply with general legislation supplemented with numerous specific legislation and regulations – therein security is implicated to a certain extent. This legislation is for the most part international and The Netherlands comply with it. On a governmental level there exists no strong drive to make national legislation. Influencing the "level playing field" by national laws is "not done" and so there exists a strong focus on Europe. This is understandable given the international position of The Netherlands as "Gateway to Europe".

The security of hazmat transport depends on the attention many stakeholders pay on the issue of security of transport and logistics. The thought is that the prevention of cargo theft will prevent the use of hazmat transporters for terrorists acts as well. Nowadays, there is relatively much effort on this security theme – by governmental incentive of by public private partnerships. Very interesting and promising prevention projects are executed. Naturally, these national projects must be extended to Europe – a security lane from the border to Rotterdam is good for The Netherlands, however, a secure lane from Rotterdam to for instance Marseille is *even better* for The Netherlands.

Bibliography

Cohen, R.E. and Ahearn, F.L. (1980) *Handbook of Mental Health Care of Disaster Victims*, Johns Hopkins University Press, Baltimore.

Herzberg, F. (1968) One more time: how do you motivate employees? *Harvard Business Review* (January–February), p. 57.

Jonkman, S.N., Vrijling, J.K., and Kok, M. (2008) Flood risk assessment in The Netherlands: a case study for dike ring South Holland. *Risk Analysis*, **28** (5), 1357–1373.

Ministerie van Economische Zaken and Directoraat-generaal Ondernemen & Innovatie, Directie Ondernemen (2009) Convenant Aanpak Criminaliteit Transportsector, Den Haag: MinEZ.

Ministerie Verkeer en Waterstaat, Rijkswaterstaat – Adviesdienst Verkeer en Vervoer, and Kennisinstituut voor Mobiliteitsbeleid (2007) Toekomstverkenning Vervoer Gevaarlijke Stoffen over De Weg 2007, Ministerie Verkeer en Waterstaat, Den Haag.

NDL/HIDC (Nederland Distributieland) and TNO Business Unit Mobiliteit en Logistiek (2009) De logistieke kracht van Nederland 2009, Zoetermeer: NDL/HIDC en TNO.

Projectgroep Monitoring (2010) Eindrapport – Convenant Aanpak Criminaliteit in de transportsector, Den Haag.

Rosenthal, U. (1985) *Disaster Management in The Netherlands: Planning for Real Events*, Erasmus Universiteit, Rotterdam.

Talbot, J. and Jakeman, M. (2009) *Security Risk Management – Body of Knowledge*, John Wiley & Sons, Inc, Hoboken, New Jersey.

Timofeeff, P. (2011) in a broadcasting on RTL TV on the 5th of January, 2011.

Visser, J. and Hendrickx, H. (2000) *Veiligheidsbeleving Goederenvervoer in Nederland in Risico's Van Het Verkeer En Vervoer: De Beleving Van De Burger*, Kennisinstituut voor Mobiliteitsbeleid, Den Haag.

WODC (2010) Monitor Criminaliteit Bedrijfsleven 2009. Feiten en trends inzake aard en omvang van criminaliteit in het bedrijfsleven, Den Haag: 2010.

Internet Sources

www.brabant.criminee.nl

http://www.businessdictionary.com/definition/security.html

http://www.businessdictionary.com/definition/safety.html

http://www.cbs.nl/nl-NL/menu/themas/bevolking/cijfers/extra/bevolkingsteller.htm

www.nctb.nl

http://www.rijksoverheid.nl/onderwerpen/vervoer-gevaarlijke-stoffen/wetgeving-en-beleid-voor-het-vervoer-van-gevaarlijke-stoffen

www.stavc.nl

14
Safeguarding Hazmat Shipments in the US: Policies and Challenges

Joseph S. Szyliowicz

14.1
Introduction

The dangers posed by the transportation of hazardous materials (hazmats) have been vividly captured in the recent adventure movie, "Unstoppable". In 2001, the engineer failed to set a brake properly when he dismounted, thus permitting a freight train with two tank cars filled with a deadly chemical (molten phenol acid) to travel over 60 miles across North West Ohio at speeds that exceeded 40 miles per hour. Finally, three railroad employees managed to catch up with the unaccompanied train and brought it to a stop. Fortunately this incident, unlike some others, ended without causing a major catastrophe.

The transportation of such materials has always involved risks and, since huge quantities of hazmats (it is estimated that up to 500 000 products can be classified as hazardous materials) are transported daily by truck and rail to and from an estimated 4.5 million sites across the US. Thus, no part of the country is immune from danger. According to the most recent data (2003) every state has endured an incident of some sort. These ranged from a low of six in Alaska to a high of 1242 in Illinois, for a total of over 15 000 incidents[1].

14.2
Intermodalism

Ensuring that the risks inevitably associated with the transport of hazardous materials be minimized is complicated by the fact that each mode – air cargo, waterborne, rail and highways – poses special challenges because of its unique characteristics. The bulk of hazmat transportation, however, takes place on land, either by rail or truck. However, a modal focus is no longer adequate in today's globalized world where freight of all kinds, including hazmats, travels

1) http://www.statemaster.com/graph/hea_haz_mat_inc_tot_num-hazardous-materials-incidents-total-number.

Security Aspects of Uni- and Multimodal Hazmat Transportation Systems, First Edition. Edited by Genserik L.L. Reniers, Luca Zamparini.

intermodally. This continually adds another potential source of danger since an ever growing number of containers (many of which enter the US by land from Mexico or Canada or by sea from numerous other countries) is hauled by both truck and rail. In 2007 there were a total of 9.2 million intermodal rail shipments. (TSA, 2007, A106).

Shipments from abroad are of particular concern because they can be used to smuggle materials, including radioactive wastes and other mass destruction materials into the US. Preventing such an event is a major focus of the US homeland security effort. However, the size and scope of the US maritime trade is enormous. It accounts for 80% of world trade by value and 90% by volume, imports cargo from thousands of ports around the world, directly or indirectly into over 360 ports. (Harrald, 2005, p. 161; Department of Homeland Security, 2008, p. 4). Seaports are traditionally located next to and often surrounded by urban areas and were designed, especially on the sea side to allow easy access for inbound and outbound ships, thus rendering them vulnerable to terrorist attacks (Small Vessel Security Strategy April 2008 available at http://www.dhs.gov/xlibrary/assets/small-vessel-security-strategy.pdf). Of particular concern are the huge numbers of containers that move through US ports. On a typical day in 2008, US container ports handled an average of 77 000 TEUs (one 20-foot container equals one TEU, and one 40-foot container equals two TEUs.) Any one of these containers could be smuggling hazardous materials ranging from nuclear weapons to the components for a "dirty bomb" not only into the country but to another country as well.

14.3
The Pre-9/11 Situation

Prior to 9/11, the private sector was focused on how to create a transportation system that could effectively and efficiently handle the new demands that were being placed upon it by ever-integrating economic systems. Everyone understood that the transport of hazardous materials created special issues but the focus was not upon security but upon safety. Two kinds of events were of particular concern. The first involved criminal activity. The railroads and the trucking companies defined security largely in terms of petty crime such as thefts of hazardous and other materials from railroad yards or of a tanker or a container. The second involved accidents (when a train or a truck was involved in a collision of some sort or when a spill occurred). The latter, in particular raised issues of great concern to state and local governments such as how to route trains and trucks carrying hazardous materials so as to bypass highly populated areas, provide advance notification to public-safety personnel, ensure that first responders were adequately trained, and that emergency communication plans had been prepared.

The national government, for its part, sought to minimize the probability that an accident might occur. Accordingly, it enacted various rules and regulations that were designed to enhance safety. The Hazardous Materials Transportation Uniform Safety Act of 1990, (HMTUSA), for example, emphasized the need to assess the risks and benefits associated with the transportation of hazardous materials by truck and rail and attempted to simplify the maze of conflicting state, local, and federal regulations. Also, in 1996 the Federal Highway Administration (FHWA) issued guidelines for the routing of trucks carrying hazardous materials that required carriers to select routes on the basis of such criteria as the level of risk, the comparison of alternative routes and also provided for public input. Routes were either "designated", those on which hazardous materials could be transported or "restricted" routes that could not be used for such transport (National Highway Institute (NHI) and Federal Highway Administration (FHWA), 1996).

However, these regulations were never fully implemented; by 1998, only a third of the states had designated routes because of the costs involved in obtaining reliable data for route assessments. (Shaver and Kaiser, 1998). Furthermore, little attention was paid to tracking and monitoring shipments, activities that the trucking companies considered too costly. And, though emergency responders would benefit from information that could be provided by the use of available new technologies, little was done to integrate ITS applications into state and local emergency-management programs (Lindquist and Slack, 1999).

Nevertheless, the Department of Transportation (USDOT) continued to strive to improve the safety of hazmat transportation and established an Hazmat Program Evaluation Team in its 1997–2002 strategic plan to assess the effectiveness of the Department's efforts. This team discharged its functions admirably. It found that the Department was not carrying out its responsibilities very effectively owing to significant weaknesses in its hazardous-materials programs, including a lack of reliable data on which to base policy and a failure to deliver its programs effectively. Perhaps this was inevitable since there was little coordination, direction and strategic planning involving hazardous-materials transportation (USDOT, 2000; Department Wide Program Evaluation of the Hazardous Materials Transportation Programs Final Report, March 2000, available at: http://www.phmsa.dot.gov/staticfiles/PHMSA/DownloadableFiles/Files/hmpe_report.pdf).

The problem of safeguarding seaports was recognized by President Clinton who established the "Seaport Commission" in 1999. Following a review of fifty major ports, it concluded that they were very vulnerable to a terrorist attack and identified a need for general guidelines and standards for physical, procedural and personnel security. In addition to discussing specific security measures, it also raised the basic issue of how to enhance maritime security without disrupting global commerce (CRS, 2002).

14.4 The Magnitude of the Problem

Each day, enormous quantities of hazardous materials are carried across the United States by the various modes and, intermodally as well. As noted above, large numbers of containers are imported and exported annually, containers that are hauled by rail and truck to and from seaports that are themselves vulnerable to attack. Intermodal shipments have grown rapidly rising from 0.4% in 1997 to 5% in 2007 (see Table 14.2).

The total amount of hazardous-materials transported was 2.2 million tons in 2007, a 50% increase since 1997. The largest mode is the highway where 400 000 large trucks carried about 1.2 million tons of hazardous materials in 2007 as compared to 870 000 tons a decade earlier. Although the railroads carry a smaller amount of hazmats, these include such perilous materials as fuels, fertilizers and explosives. The most dangerous cargoes, however, are chemicals, including chlorine, which is widely used in the manufacture of pharmaceuticals and to purify drinking water. About three million of the 12 million tons of chlorine produced annually in the United States are shipped in pressurized tank cars. Such chemicals are extremely dangerous because of the risk of explosion, fire, and the "Poison Inhalation Hazard (PIH)" or "Toxic Inhalation Hazard" (TIH) that gases or liquids such as chlorine and anhydrous ammonia pose to health and the environment, if released into the atmosphere. Flammable liquids are by far the most common type of hazardous materials as is evident from Table 14.1 which contains data on the different types of hazardous materials that were transported between 1997 and 2007. The other modes account for relatively small percentages. Furthermore, the amount carried intermodally is still quite small, though the amount carried has experienced a significant increase, rising, between 1997 and 2007, from 1.5 to 5.4% of the total. This increase has involved the longer trips for here, intermodalism increased dramatically, accounting for 1.9% in 1997 and 13.7% of the total in 2007 (see Tables 14.2 and 14.3).

All transport of hazardous materials involves numerous actors. Altogether there are 559 separate railroads, seven of which are defined as "Class 1" (AAR, 2008). These companies are but a small, albeit influential, part of the private stakeholders involved in the land transport of hazmats. Their total has been estimated as follows: shippers (45 000 regular and another 30 000 occasional ones), carriers (45 000 and 500 000 occasional ones) and a huge number of receivers. Nor can one overlook the seven powerful industry associations. The number of public actors is also significant – 19 government agencies at the national level as well as numerous state and local units are involved in one way or another, in the transportation of hazardous materials (NRC/TRB, 2005, Cooperative Research for Hazardous materials Transportation, TRB special report 283, Washington, DC). Under these conditions, it should not be surprising that despite all the structural and regulatory reforms that have been implemented in recent years in an effort to secure the transportation of hazmats from terrorist attack, controversies are common and various problems such as coordination and implementation have not yet been fully overcome.

14.5 The Impact of 9/11

The tragic events of 9/11 elevated security concerns to new heights because of the obvious possibility that terrorists might use hazardous materials to launch a devastating attack within the country. The need for action was highlighted by the

Table 14.1 Hazardous material shipment characteristics by Hazard class: 2007, 2002 and 1997.

Estimates are based on data from the 2007 and 2002 Commodity Flow Surveys. Because of rounding, estimates may not be additive.

Hazard class and description	Value			Tons			Ton-miles[1]			Average miles per shipment		
	2007 (million $)	2002 (million $)	1997 (million $)	2007 (thousands)	2002 (thousands)	1997 (thousands)	2007 (millions)	2002 (millions)	1997 (millions)	2007	2002	1997
Total	1 448 218	660 181	466 407	2 231 133	2 191 519	1 565 196	323 457	326 727	263 809	96	136	113
Class 1, explosives	11 754	7901	4342	3047	5000	1517	911	1568	S	738	651	549
Class 2, gases	131 810	73 932	40 884	250 506	213 358	115 021	55 260	37 262	21 842	51	95	66
Class 3, flammable liquids	1 170 455	490 238	335 619	1 752 814	1 788 986	1 264 281	181 615	218 574	159 979	91	106	73
Class 4, flammable solids	4067	6566	3898	20 408	11 300	11 804	5547	4391	9618	309	158	838
Class 5, oxidizers and organic peroxides	6695	5471	4485	14 959	12 670	9239	7024	4221	4471	361	407	193
Class 6, toxic (poison)	21 198	8275	10 086	11 270	8459	6366	5667	4254	2824	467	626	402
Class 7, radioactive materials	20 633	5850	2722	515	57	87	37	44	48	S	S	445
Class 8, corrosive materials	51 475	38 324	40 423	114 441	90 671	91 584	44 395	36 260	41 161	208	301	201
Class 9, miscellaneous dangerous goods	30 131	23 625	23 946	63 173	61 018	65 317	23 002	20 153	22 727	484	368	323

KEY: S = Estimate does not meet publication standards because of high sampling variability or poor response quality.
1) Ton-miles estimates are based on estimated distances traveled along a modeled transportation network.
Notes: Value-of-shipment estimates are reported in current prices. Estimated measures of sampling variability for each estimate known as coefficients of variation (CV) are also provided in these tables. More information on sampling error, confidentiality protection, nonsampling error, sample design, and definitions may be found at http://www.bts.gov/cfs.
Source: U.S. Department of Transportation, Research and Innovative Technology Administration, Bureau of Transportation Statistics and U.S. Department of Commerce, U.S. Census Bureau, 2007 Economic Census: Transportation Commodity Flow Survey, December 2009.

Table 14.2 Hazardous material shipment characteristics by mode of transportation: 1997, 2007 and 2002.

Estimates are based on data from the 2007 and 2002 Commodity Flow Surveys. Because of rounding, estimates may not be additive.

Mode of transportation	Value			Tons			Ton-miles[1]			Average miles per shipment		
	2007 (million $)	2002 (million $)	1997 (million $)	2007 (thousands)	2002 (thousands)	1997 (thousands)	2007 (millions)	2002 (millions)	1997 (million $)	2007	2002	1997
All modes	1 448 218	660 181	466 407	2 231 133	2 191 519	1 565 196	323 457	326 727	263 809	96	136	113
Single modes	1 370 615	644 489	452 727	2 111 622	2 158 533	1 541 716	279 105	311 897	258 912	65	105	95
Truck[2]	837 074	419 630	298 173	1 202 825	1 159 514	869 196	103 997	110 163	74 939	59	86	73
For-hire truck	358 792	189 803	134 308	495 077	449 503	338 383	63 288	65 112	45 234	214	285	260
Private truck	478 282	226 660	160 693	707 748	702 186	522 666	40 709	44 087	28 847	32	38	35
Rail	69 213	31 339	33 340	129 743	109 369	96 626	92 169	72 087	74 711	578	695	853
Water	69 186	46 856	26 951	149 794	228 197	743 152	37 064	70 649	68 212	383	S	S
Air (includes truck and air)	1735	1643	8558	S	64	66	S	85	95	1095	2080	1462
Pipeline[3]	393 408	145 021	85 706	628 905	661 390	432 075	S	S	S	S	S	S
Multiple modes	**71 069**	**9631**	**5735**	**111 022**	**18 745**	**6022**	**42 886**	**12 488**	**3081**	**834**	**849**	**645**
Parcel, U.S.P.S. or courier	7675	4268	2874	236	245	143	151	119	78	836	837	697
Parcel, U.S.P.S. or courier	7052	–	–	11 706	–	–	10 120	–	–	779	–	–
Truck and water	23 451	–	–	36 588	–	–	12 380	–	–	1010	–	–
Rail and water	5153	–	–	5742	–	–	2937	–	–	1506	–	–
Other multiple modes[4]	27 739	5363	2861	56 750	18 500	5879	17 297	12 369	2982	233	1371	S
Other and unknown modes	**6534**	**6061**	**7945**	**8489**	**14 241**	**17 459**	**1466**	**2342**	**1837**	**58**	**57**	**38**

KEY: S = Estimate does not meet publication standards because of high sampling variability or poor response quality. – = Zero or Less than half the unit shown; thus, it has been rounded to zero. X = Not applicable or comparison data unavailable.

1) Ton-miles estimates are based on estimated distances traveled along a modeled transportation network.
2) "Truck" as a single mode includes shipments that were made by only private truck, only for-hire truck, or a combination of private and for-hire truck.
3) Estimates for pipeline exclude shipments of crude petroleum.
4) The 2007 and 2002 "Other multiple modes" categories are not directly comparable due to a definition change. For 2002, "Other multiple modes" includes shipments using "Trucks and rail", "Truck and water", "Rail and water", and other mode combinations not specifically listed. For 2007, "Truck and rail", "Truck and water", and "Rail and water" are not part of "Other multiple modes".

Notes: Value-of-shipment estimates are reported in current prices. Estimated measures of sampling variability for each estimate known as coefficients of variation (CV) are also provided in these tables. More information on sampling error, confidentiality protection, nonsampling error, sample design, and definitions may be found at http://www.bts.gov/cfs.

Source: U.S. Department of Transportation, Research and Innovative Technology Administration, Bureau of Transportation Statistics and U.S. Department of Commerce, U.S. Census Bureau, 2007 Economic Census: Transportation Commodity Flow Survey, December 2009.

Table 14.3 Hazardous material shipment characteristics by mode of transportation: per cent of total for 1997, 2007 and 2002.

Estimates are based on data from the 2007 and 2002 Commodity Flow Surveys. Because of rounding, estimates may not be additive.

Mode of transportation	Value (per cent)			Tons (per cent)			Ton-miles (per cent)[1]		
	2007	**2002**	**1997**	**2007**	**2002**	**1997**	**2007**	**2002**	**1997**
All modes	100	100	100	100	100	100	100	100	100
Single modes	94.6	97.6	97.1	94.6	98.5	98.5	86.3	95.5	98.1
Truck[2]	57.8	63.6	63.9	53.9	52.9	55.6	32.2	33.7	28.4
For-hire truck	24.8	28.8	28.8	22.2	20.5	21.5	19.6	19.9	17.1
Private truck	33.0	34.3	34.5	31.7	32.0	33.4	12.6	13.5	10.9
Rail	4.8	4.7	7.1	5.8	5.0	6.2	28.5	22.1	28.3
Water	4.8	7.1	5.8	6.7	10.4	9.1	11.5	21.6	25.9
Air (incl truck and air)	0.1	0.2	1.8	S	–	–	S	–	–
Pipeline[3]	27.2	22.0	18.4	28.2	30.2	27.6	S	S	S
Multiple modes	**4.9**	**1.5**	**1.2**	**5.0**	**0.9**	**0.4**	**13.3**	**3.8**	**1.2**
Parcel, U.S.P.S. or courier	0.5	0.6	0.6	–	–	–	–	–	–
Truck and rail	0.5	–	–	0.5	–	–	3.1	–	–
Truck and water	1.6	–	–	1.6	–	–	3.8	–	–
Rail and water	0.4	–	–	0.3	–	–	0.9	–	–
Other multiple modes[4]	1.9	0.8	0.6	2.5	0.8	0.4	5.3	3.8	1.1
Other and unknown modes	**0.5**	**0.9**	**1.7**	**0.4**	**0.6**	**1.1**	**0.5**	**0.7**	**0.7**

KEY: S = Estimate does not meet publication standards because of high sampling variability or poor response quality. – = Zero or less than half the unit shown; thus, it has been rounded to zero.

1) Ton-miles estimates are based on estimated distances traveled along a modeled transportation network.
2) "Truck" as a single mode includes shipments that were made by only private truck, only for-hire truck, or a combination of private and for-hire truck.
3) Estimates for pipeline exclude shipments of crude petroleum.
4) The 2007 and 2002 "Other multiple modes" categories are not directly comparable due to a definition change. For 2002, "Other multiple modes" includes shipments using "Trucks and rail", "Truck and water", "Rail and water", and other mode combinations not specifically listed. For 2007, "Truck and rail", "Truck and water", and "Rail and water" are not part of "Other multiple modes".

Notes: Value-of-shipment estimates are reported in current prices. More information on sampling error, confidentiality protection, nonsampling error, sample design, and definitions may be found at http://www.bts.gov/cfs.

Source: U.S. Department of Transportation, Research and Innovative Technology Administration, Bureau of Transportation Statistics and U.S. Department of Commerce, U.S. Census Bureau, 2007 Economic Census: Transportation Commodity Flow Survey, December 2009.

release of an FBI threat assessment in 2002 that claimed that al-Qaeda was planning to derail or detonate hazmat tank cars in urban areas (Bogdanich and Drew, 2005). Accordingly, the government initiated a number of important measures in an effort to safeguard the hazmat transportation systems but such factors as the nature of the American system of government, the number and diversity of stakeholders and actors, the size of the transportation infrastructure and the quantity of the shipments all presented–and continue to present–significant obstacles to developing and implementing policies and projects to effectively safeguard hazmat shipments.

The USDOT, which has traditionally been responsible for transportation safety and security, is the lead agency for safeguarding hazmat shipments. The Homeland Security Act (2002) explicitly accorded the Secretary of Transportation that responsibility. Within the DOT, the Pipeline and Hazardous Material Safety Administration (PHMSA), created in 2004, is the unit charged with the overall regulation of the movement of hazardous materials. To enhance security, the PHMSA has issued (and continues to issue) a large number of regulations that establish criteria for classifying, packing, handling, and transporting hazardous materials and communicating and reporting issues and incidents.

It is not clear, however, whether the PHMSA enjoys adequate resources to carry out this function effectively. Its FY 2009 request totals $168 million, a marked increase from its 2007 budget of $134 million. However, one of its primary responsibilities is the safety of pipelines (pipelines, though not their pumping stations, are generally considered to be low-salience terrorist targets) and most of its budget ($93 million) is allocated to that function. Even though that figure represents a sharp increase from $75 million in 2007, questions have been raised whether even these staffing levels are adequate for pipeline safety inspections and enforcement (CRS, 2010). Another big increase, from $14 million to $28 million is evident in the Emergency Responders Grant Program which allocates money to state and local governments for training and equipment. Funding for hazmat security and safety, however, has remained essentially stable, $27 million in 2007, $28 million in 2009[2]. Funding antiterrorism programs involves difficult choices but many observers believe that the allocation of resources has been too heavily biased towards the aviation sector, which has received a disproportionate amount of the available resources, at the expense of other modes and dangers.

The regulations issued by the PHMSA are enforced by the modal agencies such as the Federal Railroad Administration (FRA), the Federal Aviation Administration (FAA), and the Federal Highway Administration (FHWA) as well as the Coastguard. Railroad security is the domain of the FRA, which works closely with the Association of American Railroads (AAR). It shares intelligence with the railroad companies, conducts training for first responders, and inspects shipments and railroads and their infrastructures such as bridges and tracks to ensure compliance with the DOT's regulations. Highway transport of hazmats is overseen by the Federal Motor Carrier Safety Administration (FMCSA) whose agents work with

2) http://www.dot.gov/bib2009/htm/PHMSA.html.

thousands of trucking companies to ensure regulatory compliance. It requires almost all individuals and companies in the hazmat industry to implement a detailed security plan and to ensure that all personnel receive appropriate training. It has also developed a "Risk Management Self Evaluation Framework" to help companies develop their plans. It has conducted numerous reviews of motor carriers' security plans, provided grants to border states to enhance inspection of inbound cargo, worked to develop various relevant technologies, and provided intelligence alerts to drivers and companies.

All these activities are supplemented by those of the Department of Homeland Security (DHS), which was established after 9/11 in order to create a centralized agency to protect against terrorist attacks. As such, it is responsible for safeguarding the country's critical infrastructure, including transportation. It determines threat levels, carries out risk assessments, issues advisories, develops security strategies, and operates an extensive research and development effort to deploy new technologies. Within the DHS the Transportation Security Administration (TSA) is responsible for all aspects of transportation security. However, its primary focus has been upon the aviation sector and its hazmat-related activities have been diffuse and limited. Responsibility for hazmats is dispersed among a number of offices within the TSA and its hazmats-related programs have been largely limited to information and training and to checking the drivers of hazmat trucks against various data bases to ensure that they are not security risks. The efficacy of these activities has been questioned and it is generally believed that they have, at best a "limited impact" (CRS, 2005a, p. 14). Indeed, the TSA has honestly admitted that it ". . . has been unable to place the focus on hazmat transportation security that is warranted" (CRS, 2005a, p. 8).

Under these conditions, it is not difficult to conclude that the PHMSA's and the TSA's efforts undertaken after 9/11 did not greatly reduce the vulnerability of hazmat rail transport. This was brought home tragically on January 6, 2005 when 2 trains collided in Granitcvillc (SC). The locomotives and 17 freight cars, derailed, one of which, loaded with 90 tons of chlorine ruptured, releasing large amounts of the gas. Nine people died, over 250 people were treated for chlorine exposure, and more than 5000 persons had to be evacuated. This disaster provoked an immediate reaction. The FRA conducted a thorough investigation and within four months launched the National Rail Safety Action Plan. In cooperation with the railroads, it developed programs to deal with the two primary causes of accidents, those caused by human factors because of such factors as fatigue and those caused by defective tracks. The FRA's compliance program was strengthened, measures were taken to improve the safety of highway–rail crossings (an obvious source of attack), safety measures for hazardous materials were upgraded and emergency preparedness enhanced. Attention was paid to developing technologies for non-signaled tracks and for enhanced rail tank car structural integrity and for ensuring that emergency responders had accurate information concerning hazardous materials rail shipments (USDOT, 2008).

The basic strategy underlying hazmat security also changed significantly. It evolved into a "layered approach" that incorporated security measures at all stages of a hazmat shipment whether by road or rail. All employees dealing with hazmats were now to be subject to reliable background checks and certification that they had undergone appropriate training and all transportation operations, including routing would be carried out according to specific plans and procedures based on detailed risk assessments. These requirements would be checked by federal and state inspectors (CRS, 2006, Transportation Issues for the 110th Congress). The effectiveness of this new strategy depends, of course, on how it has been implemented by the railroads and the truckers.

Similar concerns apply to efforts to secure global supply chains that also received immediate attention. In 2002, Congress passed the Maritime Transportation Security Act (MTSA) that specified various initiatives to enhance port and vessel security. The lead agency is Customs and Border Protection (CBP) now part of the Department of Homeland Security. It launched the Container Security Initiative (CSI) shortly after 9/11. It seeks to identify high-risk containers before they leave for the US. The number of foreign governments who participate has grown from eleven in November 2002 to 58 today. However, moving inspections overseas raises questions about the integrity, reliability, and honesty of the local officials who are working with US customs officials to screen the containers and about the accuracy of the manifests that have to be submitted 24 h prior to the container's departure. Closely related is the Customs-Trade Partnership Against Terrorism (C-TPAT) program that encourages the private sector to meet certain standards and to apply them from port to port. Another program, the Ten + Two has recently been introduced to provide an additional layer of protection. Importers are required to provide ten items of information, the carrier, another two. However, as in the other programs, the reliability of the data is the critical issue (Giermanski, 2011).

14.6 The Rail Sector

14.6.1 Vulnerabilities

Today, there is general agreement that rail is a safer mode of transportation for hazardous materials than highways. The latest available data shows that the number of incidents involving hazmats was, during the period from 2001–2010, by far the greatest for the highway mode, roughly 20 times the number for the rail mode, ten times the number of air transport and 300 times greater than the water mode, see Table 14.4.

According to the AAR, 99.996 per cent of rail hazmat shipments (over 1.5 million carloads in 2007) reach their destination without a release caused by a train accident.

However, rail's vulnerabilities should not be minimized. These include the size of the system and the fact that railroads are private entities who own and control

Table 14.4 Hazardous materials incidents by mode and year.

Mode of transportation	2001	2002	2003	2004	2005	2006	2007	2008	2009	2010	Grand Total
FAA-AIR	1083	732	750	993	1655	2408	1556	1278	1358	1292	**13105**
FMCSA-HIGHWAY	15804	13502	13594	13069	13460	17160	16933	14808	12731	12613	**143674**
FRA-RAILWAY	899	870	802	765	745	703	752	750	643	749	**7678**
USCG-WATER	6	10	10	17	69	68	61	99	90	105	**535**
Grand Total	**17792**	**15114**	**15156**	**14844**	**15929**	**20339**	**19302**	**16935**	**14822**	**14759**	**164992**

Source: Hazmat Intelligence Portal, U.S. Department of Transportation. Data as of 3/31/2011.

and operate almost 140 000 miles of track, the "Class 1" railroads, 95 000 miles, the others, another 45 000 miles (AAR, 2008). To this extensive network must be added an extensive infrastructure of bridges, tunnels, signals and sidings. Moreover, the much larger capacity of tank cars has the potential to create much more serious repercussions if its contents were to be released and it is more difficult to avoid routing trains through high-risk urban areas.

Terrorists can exploit these vulnerabilities in many ways. They could attack the track infrastructure using a simple element like a crowbar, though some knowledge of physics is required to change the gage sufficiently to derail a train or reposition switches. Explosives could be used to destroy a tunnel, culvert, or bridge while a train is passing but this is more difficult owing to the need to place the explosives and time their detonation precisely. Still more difficult would be an attack on the signal system. Terrorists could also gain access to a rail yard where tank cars loaded with dangerous chemicals are waiting and arrange to rupture the container as it travels though the city (Hartong, Goel, Wijesekera, 2008, pp. 18–20).

Doing so would permit the deadly gas to rush out in a few minutes and create a toxic cloud that would rival the impact of a WMD. Fortunately, this scenario falls under the "high-impact low-probability" category because it would require a high level of skill and practice, and is difficult to implement. A second line of attack is, however, easier to execute. This involves creating a small rupture, using a weapon or drilling a small hole. The deadly gas would escape very slowly along the train's route and diffuse to the areas around the tracks and beyond. In an urban area, thousands would inhale the poison before the source was identified and steps taken to deal with the problem. The extent to which medical facilities would be able to deal with such mass casualties is questionable, given the kinds of equipment required (Coyle, 2009a).

14.6.2
Solutions and Their Effectiveness

Being fully aware of such potentially deadly consequences, the U.S. Department of Transportation's Pipeline and Hazardous Materials Safety Administration (PHMSA) and the FRA have initiated a number of important measures, as has the AAR that created a special committee to deal with hazardous cargoes. That committee has established special procedures for trains carrying such freight and implemented a Terrorism Risk Analysis and Security Management Plan, in cooperation with the DHS. It has worked with communities to prepare emergency-response plans and to train emergency responders and cooperates with the Chemical industry's Chemical Transportation Emergency Center that acts as a central communications node for first responders. It has also encouraged the development and deployment of new technologies such as train control systems, electronic pneumatic brakes and communication and tracking systems. That committee has also worked to improve the ability of chlorine and anhydrous ammonia tank cars to withstand a rupture (AAR, 2007).

These voluntary efforts, however, did not yield what were deemed adequate reinforced tank car standards and in 2009 the PHMSA issued a formal rule doubling the strength of tank cars carrying Poison Inhalation Hazard (PIH) commodities so as to minimize the possibility of leakages. Tank cars had to withstand 25 mph for side impacts and 30 mph for head-on collisions – more than double the existing requirement. To address the concern that PIH tank cars manufactured prior to 1989 might not be adequately resistant, older tank cars were to be phased out. Furthermore, trains with such a PIH tank car coud not exceed 50 mph and trains with older PIH cars not meeting the new standard or that were traveling in nonsignaled areas were not to exceed 30 mph (74 FR 1769 – Final Rule, January 13, 2009 available at: http://www.phmsa.dot.gov/portal/site/PHMSA/menuitem.ebdc7a8a7e39f2e55cf2031050248a0c/?vgnextoid=4e007ee0063ce110VgnVCM1000001ecb7898RCRD&vgnextchannel=26a1d95c4d037110VgnVCM1000009ed07898RCRD&vgnextfmt=print). The railroads and the chemical industry, however, argued successfully that these requirements were too stringent, would hinder service, and imposed unnecessary burdens. Their efforts led to a significant weakening of the proposed rule. The tank-car standards were replaced by commodity specific requirements, the speed limit was abandoned, and the new standards would be applied incrementally (Branscomb *et al.*, 2010, pp. 38–40).

The dangers posed to urban areas could, of course, be minimized by not sending through shipments into major cities. Routing decisions, however, have traditionally been the sole purview of the railroads; they have selected the most desirable route to carry these shipments from origin to destination on the basis of their own cost–benefit calculations. This situation was challenged in 2005 by Washington D.C. and, subsequently other cities also sought to force the railroads to select alternate routes. In order to deal with this problem, new regulations were published in 2008 that required railroads to carry out an annual analysis of their routes in order to identify those that pose the least risk, using 27 criteria including population and traffic density, potential threats, and emergency-response capability. The FRA reviews the analysis and can require a railroad to choose an alternate route until the identified weaknesses are remedied (Federal Railroad Administration Rail Hazmat Routing Rule Fact Sheet; available at: http://www.fra.dot.gov/Downloads/FRA%20Rail%20Hazmat%20Routing%20Rule%20Fact%20Sheet%20(December%202008).pdf).

These regulations have been widely condemned for many reasons. First, the railroads retain the right to select the routes; state and local officials play no role in these decisions, even though the railroads are expected to work with them. Secondly, the railroads are the ones who decide how much importance to accord to each of the 27 variables (which are unranked by the DOT) that they must consider in carrying out the risk assessment. Thirdly, only those with "a need to know," are allowed access to the analyses and route selections. The public is explicitly excluded because of obvious security concerns. Fourthly, no federal approval of the selected route is required although the FRA has announced that it will evaluate the routing decisions. That this represents a significant oversight is extremely doubtful given the number of FRA inspectors and the extensive calculations involved. Moreover, no explicit evaluative standards or guidelines on how to weight the 27 variables have been announced. Congress would have to significantly expand the inspection force employed by the FRA to give that agency any chance of effectively policing this routing rule. Finally, even if a railroad identifies an alternative route as safer and more secure, it may continue to use the one through the target city by implementing some unspecified "remediation and mitigation" measures[3].

These shortcomings are obviously due to the AAR's opposition to mandatory routing. It argues that doing so creates new vulnerabilities because it "merely shifts the risk from one community to another . . . (and) hazmat spends more time on the rails . . ."[4]. However, while it is true that smaller communities with more limited resources would now be vulnerable, routes that bypass high-threat urban areas (HTUA) would reduce the size of the exposed population. Also, it is not obvious that lengthened travel time equates to higher risk. A key issue is, of course the financial consideration. Rerouting would inevitably involve higher costs. The AAR is, quite naturally, concerned with financial considerations and, therefore, also seeks to repeal the government's ability to compel railroads to carry TIH materials because of the potential liabilities that the railroad would incur in the case of an incident.

Additional measures are designed to deal with various operational vulnerabilities concerning rail yards and sidings where tank cars often sit unguarded for various periods of time. Traditional problems include the number of workers who have access and the level of their training (Allen and Fronzcak, 2007). Now, the railroads must work to minimize delays and storage and carry out enhanced

3) Friends of the Earth Blasts Bush Admin Railroad Routing Regs, http://www.foe.org/friends-earth-blasts-bush-admin-railroad-routing-regs.

4) Association of American Railroads (AAR), HAZMAT-FAQ. Available at: http://www.aar.org/Safety/Hazmat/Hazmat-FAQ.aspx.

pretrip inspection in order to reduce the possibility that an explosive device has been inserted. (AAR web site) Employers must train all employees dealing with hazmats and retrain them every three years, classify and identify the hazardous materials according to specific criteria, package the materials appropriately, provide hazard warning information though shipping papers, labels, placards, emergency contacts, and report all incidents (PHMSA web site).

Much obviously depends on how effectively the railroads implement these security initiatives and history is not comforting. For instance, a former employee of the FRA sent an inspector to a Las Vegas rail yard following a threat report. He reported that the security measures "were virtually nonexistent . . . he found no one watching over six tank cars with markings indicating that they might contain chlorine gas. . . . two hours later he visited another rail yard with four tank cars possibly carrying poisonous gas and they, too, were unguarded." Subsequently, the AAR announced that the railroads police force had implemented over fifty new measures to better control rail yards and sidings (Bogdanich and Drew, 2005). Hopefully these measures have yielded positive results.

Nor are the railroads required to tell jurisdictions the nature of their cargoes although it is "recommended that they do". In practice, this is rarely done so that municipalities through which railroads carry their dangerous cargoes are provided with very little if any information about the kinds and amounts of hazardous chemicals that move through their boundaries. The only information that is available to them are the placards that are attached to the railcars carrying dangerous materials. These mandatory signs are designed primarily to inform emergency responders who have to deal with an incident; however, they are of little use to communities seeking to develop emergency-response plans (Coyle, 2010a).

14.6.3 Emergency Planning and Response

The difficulties that a community encounters when attempting to deal with emergency planning is exemplified by the case of Denver, Colorado, the site of the 2008 Democratic National Convention that nominated Barak Obama for President. Close by were railroad tracks, a situation that raised the obvious issue of the potential risks to the delegates and high officials posed by the rail transport of hazardous materials. Officials responsible for security planning raised this point early. The railroads, however, were reluctant to cooperate in any manner, refusing to identify or reroute the cargoes that were being carried so close to where the meetings were taking place and to the stadium that was expected to be filled with people who wished to hear Mr Obama's acceptance speech. Appeals to the railroads for informationand action by the FBI and state and local officials were in vain and only when the highest levels of the federal government became involved did the railroads agree to cooperate.

This situation raises two fundamental issues. The first involves the first responders who have to deal with an incident. They have to rely on those mandated signs and upon calling CHEMTREC, a centralized clearing house established by the

American Chemistry Council to provide information about particular hazmats. Numerous training manuals, including an Emergency Response Guidebook, and courses have been developed for first responders and a DOT grant program is designed to improve planning and training. However, questions continue to be raised as to whether first responders can be adequately prepared to deal with an incident if they are not informed, in advance, of the kinds of chemicals being shipped through their communities.

One expert has suggested a simple solution to this problem–making the data concerning TIH shipments (the routes regularly involved and the chemicals that are being carried) available to the State Fusion Centers. These already function as nodes for information sharing between federal, state, local, and private party stakeholders concerning threat and related intelligence and could pass the information on to the Local Emergency Planning Committees (LEPC) whose members include state and local officials, police, fire, and public health professionals and that are responsible for developing an emergency response plan (ERP) for the community. The data provided by the Fusion Centers could then be incorporated into the ERP. Moreover, existing technology could be deployed on trains and such details as the location of each TIH car within the train could be transmitted to responders controlled by emergency response agencies (Coyle, 2010a).

Such measures would be very helpful but still more needs to be done to ensure the safety and effectiveness of emergency personnel for they lack some essential equipment. A 2008 GAO study summarized the problem as follows: "local first responders still do not have tools to accurately identify right away what, when, where, and how much chemical, biological, radiological, or nuclear (CBRN) materials are released in US urban areas, accidentally or by terrorists" (GAO, 2008a).

14.6.4
The Federal Government and the Private Sector

The second involves the difficulty of striking a balance between implementing security measures without adversely impacting commerce, of how to resolve conflicts when government and/or community interests come up against the interests of the private sector who are the owners and operators of the railroads and its infrastructure. As the case cited above suggests, the railroads (and other powerful private-sector actors such as the chemical industry) enjoy a high degree of political power, are able to defend their commercial interests effectively and can be overridden only with difficulty.

Nevertheless, some aspects of hazmat security have been enhanced by rules and regulations that have been opposed by the private sector. The TSA has issued a rule that seeks to ensure a secure "chain of custody". The railroads are required to inspect the car before it is picked up at the point of origin, have a witness and formal documentation for every transfer. Railcars containing hazmats are to be tracked in transit and this information is to be provided to the TSA when needed. The existing system of sensors on tracks takes six to eight hours to provide a railcar's location but the new rule requires the major railroads to respond within five minutes of a request, 30 min if several cars are involved, though only of the last reported location. Any suspicious activity is to be reported immediately to local law enforcement, then to the TSA. All railroads are to appoint a Rail Security Coordinator (and an alternate) who is always available to the TSA.

The TSA is also authorized to carry out surprise inspections, a measure strongly opposed by the chemical industry that argued that it would disrupt their operations and create the risk that a terrorist might impersonate an inspector. Questions were also raised about the effectiveness of these new rules and, especially their cost. The PHMSA has estimated that its routing rule would cost railroads $20 million over 20 years, the TSA that its rule would cost $22 million the first year, then decline to about $12 million in ten years. The railroads (who earn $42 billion annually) and the chemical manufacturers have argued that these rules are not cost effective. Measuring cost effectiveness is, of course, no easy matter since one can only guess at the possibility that a terrorist would attack a hazmat rail car. Nevertheless, as the PHMSA noted "the fact an event is infrequent or has never occurred does not diminish the risk or possibility of such an event occurring" (GAO, 2008a).

How much security all these efforts have achieved is difficult to assess. One chemical security expert has argued that the FRA has not paid adequate attention to the security dimension as opposed to the dangers posed by collisions and that even when dealing with security, it has not been very imaginative; terrorists are flexible and innovative but though the "TSA has looked at scheduling and hand-offs of TIH cars . . . the only kind of attack that they have addressed at all is the placement of an IED" (Coyle, 2010b, "Ballistic Protection for Hazmat Transport", Chemical Facility Security News). Nor is it possible to evaluate the effectiveness of the security measures adopted by the railroads since, as of 2008, the TSA has never implemented measures to monitor and evaluate these efforts. Without such a mechanism, as the GAO pointed out, the TSA is unable to direct its efforts to remedying identified security gaps. Furthermore, the nature of the existing patterns of cooperation and collaboration between the railroads, the AAR and the government require attention and need to be improved, as did the coordination that characterized the relationship among the government agencies dealing with railroad security (GAO, 2008b).

14.7 Highways

14.7.1 Vulnerabilities and Policies

As noted above, trucks carry a very large amount of hazardous materials. They do so over an extensive road system totaling 4 million miles, about 50 000 miles of which are interstate highways. Ownership of these roads is divided among the federal, state and local governments who own 3%, 19.5% and 77.5%, respectively[5].

This system poses significant security threats that are difficult to deal with. To achieve highway security requires resolving many of the same problems that complicate efforts to enhance railroad security as well as specific problems created by the ease with which trucks can be obtained and the well-known appeal of explosive loaded trucks as weapons. Moreover, motor and heating fuel are transported in large quantities in tank trucks that can also be used as very effective weapons to attack a variety of targets. Unfortunately, these trucks are difficult to safeguard, as is evidenced by the ease with which they can be (and are frequently) stolen.

At the national level, the Federal Motor Carrier Safety Administration (FMCSA) Hazmat Division is responsible for developing policies and procedures that ensure regulatory compliance by those carriers and shippers of hazardous materials who are required to carry out security plans. Such plans involve personnel security, facility security, *en-route* security and security awareness training for employees. The DOT Routing Regulations provides guidelines that establish mandatory requirements for states that establish or modify hazmat routes. These include the identification of local hazmat companies and determining the specific products they haul, their routes, and their security plan. It also carries out roadside inspections, works with other DOT units to develop new regulations and carries out research and analysis (www.fmcsa.dot.gov/safety-security/hazmat/hm-theme.htm).

14.7.2
Administration and Coordination

As in the case of railroads, however, these efforts are limited in their effectiveness. The quality of inspection that is carried out by both federal and state personnel is constrained by a lack of resources and necessary technologies, inadequate training, the greater difficulties presented by security as opposed to safety checks, and the prevalence of inspections on highways far away from heavily populated areas (CRS 2005a, p. 15). Such shortcomings are aggravated by coordination and related issues between and within various federal agencies, especially the DHS and the DOT. Although various risk-assessment initiatives have been launched, these were not coordinated and the results remain within the unit that carried out the analysis. Furthermore, the TSA's strategy to enhance the security of the highway infrastructure is not based on available risk assessments and information. Hence, the "DHS cannot provide reasonable assurance that its current strategy is effectively addressing security gaps, prioritizing investments based on risk, and targeting resources toward security measures that will have the greatest impact." Furthermore, the "TSA does not have a mechanism to monitor protective measures implemented for critical highway infrastructure assets. . . . Without such a monitoring

5) USDOT/FHWA, "Status of the Nation's Highways, Bridges and Transit:2004 Conditions and Performance" available at: http://www.fhwa.dot.gov/policy/2004cpr/chap2b.htm

mechanism, TSA cannot determine the level of security preparedness of the nation's critical highway infrastructure" (GAO, 2009a, 2009b). Nor is the DOT immune from criticism. One observer has gone so far as to argue that the TSA and the DOT do not "understand supply chain tracking and security, and the technology currently available . . ." (Giermanski, 2010, p. 33).

The weaknesses that characterize the federal agencies charged with safeguarding hazmat highway transport in the US, especially the PHMSA, have been highlighted by the measures specified by the Hazardous Material Transportation Safety Act of 2009. Although this legislation has not yet been enacted, it has already yielded positive results because the PHMSA has initiated a number of significant reforms. For example, the PHMSA's approach to the issuance of special permits that allowed companies to bypass the regulations has dramatically changed. Although these should have been issued with great care, the PHMSA handed them out without much attention to the recipients' previous records and even to trade associations whose member companies could thus transport hazmats without paying any attention to regulations. Nor did the PHMSA carry out any oversight of these permits or coordinate with such agencies as the FAA, FMCSA, and FRA, even though required by law (OIG, 2010, 2009).

A similar situation existed in regards to explosive classification approvals. The PHMSA lacked formal, uniform rules and though rules existed for reclassifying explosives, it did not adhere to them. Nor did it have any mechanism for resolving internal debates over safety decisions (CRS 2009a, 2009b).

PHMSA was obviously a malfunctioning organization characterized by poor managerial and administrative practices. However, it has been drastically reformed. Cynthia L. Quarterman, the PHMSA Administrator stated recently: "we had to make leadership changes not only at the highest levels but throughout the organization." One weakness, however, remains to be overcome. Funding for its hazmat division was (as noted earlier) probably inadequate (the bill called for the addition of 64 new positions) and it probably remains inadequately resourced to this day since only a handful of new inspectors have been hired (Quarterman, 2010a).

Under its new leadership, the PHMSA has also attempted to deal with some of the other issues highlighted by the 2009 bill. In regards to the unprotected product piping on cargo tank vehicles, it sought to mandate that new vehicles not transport dangerous liquids in external product piping. Although it encountered strong opposition from the ATA and the National Tank Truck Carriers (NTTC), it issued, in December 2010, a proposed rule forbidding this practice. What the final rule will look like remains to be seen though one might reasonably expect that, as in the case of the rail tank-car regulations discussed above, it will be weakened to a greater or lesser degree.

The PHMSA is also paying renewed attention to other areas highlighted by the Hazardous Material Transportation Safety Act of 2009 including data collection, analysis and dissemination, especially as they relate to emergency responders. These issues include the lack of minimum standards for hazmat emergency-response information services, the need to establish a national data and information network, obtaining better information concerning hazmat incidents and better control and training of persons involved in hazmat transportation. Through the Hazardous Materials Emergency Preparedness (HMEP) Grants Program, the PHMSA is allocating resources to better prepare them to deal with hazmat incidents. It has also established the web-based National Hazardous Materials Fusion Center to provide the latest information about such topics as lessons learned and relevant technologies (Quarterman, 2010a, 2010b).

The issue of training, however, extends beyond responders and involves all employees working with hazmat shipments. This is an important aspect of any layered system but existing efforts are not considered adequate. Even though there may be formal compliance, many companies provide only limited training to their employees (Byrd, 2009).

Closely related is the research issue. Despite its importance, this area has been largely neglected though many different agencies and private organizations carry out related research. Indeed, a recent study stated: "Perhaps the most notable gap in this system for ensuring hazardous materials safety and security is in the conduct of research" (TRB, 2005, Cooperative Research for Hazardous Materials Transportation), p. viii. The problem is that existing efforts are almost entirely oriented towards an institution's perceived needs so that little or no cooperative research is undertaken that brings together the available expertise to deal with larger issues. This need has been recognized for many years; in 1997 a special workshop attempted to establish a pilot project to demonstrate its feasibility. In 2005 another year-long study reached similar conclusions suggesting that such research would "prove useful, and be essential" (p. 2). It reiterated the previous recommendation that a pilot program be initiated. Key research needs identified by the participants included (i) data analysis for policy making and regulation, (ii) planning and preparing for emergencies, and (iii) supporting first responders (pp. 63–64). However, the weakness remains; efforts to pass relevant legislation such as the 2007 and 2009 Hazardous Materials Cooperative Research Act have failed to become law.

14.7.3
Local Governments and the Private Sector

The 2009 bill also highlighted the serious issues that are evident at the state and local government levels. The states' role in regards to trucking is different from that in regards to the railroads. There, it is more restricted because federal legislation including the Hazardous Materials Transportation Act and the Federal Railway safety act prevents or at least hinders states from enacting legislation that involves interstate commerce (CRS, 2005b). In the case of trucking, however, much of the traffic is intrastate and the states control much of the infrastructure, such as bridges and tunnels, for which they are responsible.

There is little doubt that hazmat transport by highways poses specific and dangerous issues for tunnels and bridges. A team of experts identified about 1000 (out of the 600 000 bridges) that, if attacked successfully could involve significant casualties and economic damage. There are also 548 tunnels, many of them under water, where an incident would create significant problems[6]. Accordingly, important security measures have been implemented including the use of cameras and sensors as well as rerouting. In Colorado, for example, trucks carrying hazardous materials are not permitted to transit a tunnel under the Rocky Mountains but have to detour over a mountain pass. When weather conditions prevent such a route, the tunnel is closed to all other traffic. Such measures are necessary and probably adequate for, as will be discussed below, many other attractive targets can be more easily attacked.

Thus, although their legislation may not conflict with federal law, the states have an important regulatory role but, as the 2009 bill noted, there is a lack of state level coordination and a program to develop uniformity in state registration and related procedures should be established. The problem is obvious. 50 state agencies and a large number of local agencies supplement the regulations regarding routes, packaging and procedures that are issued by the dozen or so federal agencies that are involved in decision making in regards to the highway transport of hazmats.

This situation creates two separate problems. First, achieving uniformity between states has always been difficult, despite ongoing efforts by the National Council of State Legislatures (NCSL). Not all states require hazmat carriers and vehicles to register and the regulations vary from state to state, although an attempt to establish uniform procedures has been established. The Trucking industry, concerned with the difficulties imposed by differing programs for interstate commerce encouraged the establishment of The Alliance for Hazmat Transportation Procedures in order to provide uniformity in regards to permits and registrations. However, this program does not overrule existing state programs and is voluntary. At present, only seven states are members so that it is far from providing a national set of uniform rules and regulations (http://www.ncsl.org).

The second problem is the ways in which the shared responsibilities between the national government and the states are handled. The states have serious reservations about the efficacy of the existing arrangements and have expressed their concern about such issues as the lack of "a clear delineation of regulatory authority and responsibility at each level or branch of government" of "uniform standards for technical requirements, routing and notification" and of the need to review "federal registration and permit programs for consolidation and administration by better established state programs". In their view, "The appropriate federal role should be to coordinate the national regulation of hazardous-materials transportation, provide technical guidance, and ensure adequate financial support" and it should always consult with the states when considering highway standards (NCSL, 2010–2011).

Further complicating the effort to achieve security on the highways is, as with rail transportation, the important role played by trade associations, in this case the American Trucking Association (ATA). Like the AAR, the ATA is concerned with the costs and obligations imposed by regulations. Thus, it opposed efforts by state and local municipalities to determine routing, challenging efforts by Washington,

6) http://www.globalsecurity.org/security/systems/highways.htm.

D.C. and Boston to restrict the passage of trucks carrying hazardous materials to certain routes because they "would create an unreasonable burden on interstate commerce and frustrate the safe and efficient transportation of hazardous materials throughout the United States." It has also challenged the PHMSA on various issues (the external piping issue discussed earlier), the manner in which the PHMSA wished to implement its authority to inspect cargoes suspected of containing undeclared hazmats on the grounds of safety and cost concerns (http://www.truckline.com/SearchCenter/Pages/Results.aspx?k=hazmat).

The ATA also opposes regulations that mandate the installation of real-time tracking technologies, arguing that these could be easily circumvented and that their costs exceed possible benefits. From the perspective of security and emergency response, however, communications and tracking technologies have a critical role to play. As one expert has noted (Coyle, 2009b):

The one thing that should be required for all hazmat shipment . . . (is) two-way communications capability. Because of the hazards associated with the load being carried, the driver should be able to communicate with first responders in what ever area the vehicle is located. I don't believe that the technology currently exists to be able to require long-haul drivers to directly contact local first responders at every point along their route. Satellite communications techniques certainly do exist to allow drivers to be able to contact a central dispatch facility. . . . Independently of the communications link there ought to be a separate system to track the location of hazmat trucks in real time. . . . This way local law enforcement and emergency response personnel could access data on the location of covered hazmat shipments in their jurisdiction.

14.7.4 Minimizing Trucking Vulnerabilities

Until now, the focus of the national government has been on enhancing the security of the most dangerous hazmat shipments such as TIH chemicals. However, a recent study has pointed out that although these are indeed attractive targets, attention must be paid to other trucking risks. One such risk is simply stealing or renting a truck, filling it with explosives and using it to attack a target. Indeed, since this weapon is so widely used by terrorists and there have been very few attacks on hazmat shipments, a recent GAO report raised the issue of whether the TSA's concentration on hazmat transportation was appropriate and whether more attention should be paid to the security of the overall trucking fleet (GAO, 2010).

This recommendation, however, requires careful consideration because protecting the huge number of trucks that travel the nation's highways would represent a very difficult and very expensive challenge. Furthermore, many trucks are rented and terrorists can easily obtain vehicles in this way, as in the case of the Oklahoma City bombing in 1995. Implementing effective security measures for the very large truck rental sector would also be extremely challenging. Moreover, for a truck to become a weapon, the terrorist must load it with explosives but doing so is no longer an easy matter, due to the enhanced efforts to control and monitor the sales of large quantities of such materials as fertilizers. Accordingly, despite the appeal of truckborne explosive attacks, the focus of security measures should be upon protecting the most dangerous potential weapon readily available to terrorists–gasoline tankers (Jenkins and Butterworth, 2010).

Gasoline tanker trucks are especially dangerous because they are ubiquitous, carrying fuel to thousands of gasoline stations (many in urban areas replete with attractive targets), and travel along regular routes. In addition, gasoline tankers are known to be lethal weapons, are easily stolen, as the numerous thefts indicate, and can be driven for hours to attack a specific target before the theft is likely to be discovered. Thus, Jenkins and Butterworth conclude (p. 2): "Gasoline tankers theoretically offer terrorists several operational attractions. . . . We therefore consider gasoline tankers and, to a lesser extent, propane tankers, to be the most attractive options for terrorists seeking hazardous-materials cargo".

The impact of a successful terrorist attack using such a tanker is clearly demonstrated by a 2007 incident when a gasoline tanker exploded on an elevated freeway section in the San Francisco Bay Area, causing the section to collapse onto the freeway below it. The resulting two-week long freeway closures affected over 80 000 commuting vehicles per day, and the collapse caused $90 million in damages ("Maze Collapse").[7)]

To minimize the possibility that terrorists could stage such an attack, Jenkins and Butterworth propose to deal with many of the security weaknesses discussed above. These include the need "to resolve the significant jurisdictional issues between federal state authorities, strengthen monitoring and enforcement of hazardous-materials security measures in the field . . . (and) work to implement vehicle-tracking technologies, panic alarms, and immobilization capabilities for vehicles carrying large quantities of specific hazardous materials, including gasoline" (p. 3). If these suggestions were to be implemented, the security of land transport of hazardous materials would be greatly enhanced. However, achieving such an outcome requires overcoming difficult obstacles within the policy process.

14.8
Conclusions

Despite the plethora of measures adopted by the US government and the private sector since 9/11, the security of hazardous-materials transportation by rail and road remains an area of great concern because of the vulnerabilities that continue to characterize the system. Proposals and efforts to make such shipments ever safer continue and new rules and regulations dealing with some aspect or other of hazmat transportation are promulgated almost daily. All these measures and projects have, without a doubt, yielded and continue to yield positive results but, as noted above, many challenges remain.

7) "Maze Collapse" Metropolitan Transportation Commission, http://www.mtc.ca.gov/news/info/2007/freeway_collapse.htm

Debates continue to rage over such topics as the cost effectiveness of various governmental policies and regulations, the routing of hazmat shipments through major urban conurbations, the best way to enhance coordination at the federal level and between the federal agencies and state and local governments, how to achieve uniformity between the states, and how best to develop and deploy specific technologies such as tracking and communication devices.

Hopefully, at least some of these security challenges will be met effectively in the near future; it is obviously necessary to deal with acknowledged security gaps. However, the terrorist threat evolves as new circumstances, capabilities, and opportunities dictate. Policy makers dealing with hazmat transportation security must also be flexible and sensitive to new dangers; there are no fixed solutions to what is an ever-changing challenge. Securing hazmat transportation requires constant vigilance, accurate intelligence, and a political commitment to a continuous process of improvement not only within the highway and rail modes but intermodally as well.

Bibliography

Allen, J. and Fronzcak, R. (2007) Comparing and Contrasting Highway and Rail Routing. Paper presented at the TRB annual meeting.

Association of American Railroads (AAR) (2007) Hazmat Transport by Rail, Washington DC.

Association of American Railroads (AAR) (2008) Overview of America's Freight Railroads, May, http://www.aar.org/PubCommon/Documents/AboutTheIndustry/Overview.pdf (accessed November 18, 2011).

Bogdanich, W. and Drew, C. (2005) Deadly leak underscores concerns about rail safety. The New York Times, January 9.

Branscomb, L. *et al* (2010) Rail Transportation of Toxic Inhalation Hazards, Harvard Kennedy School.

Byrd, L.M. (2009) Testimony before the Committee on Transportation and Infrastructure's Subcommittee on Railroads, Pipelines, and Hazardous Materials, November 16.

Congressional Research Service (CRS) (2002) Maritime Security: Overview of Issues, July 19.

Congressional Research Service (CRS) (2005a) Rothberg, Paul F., Security Threat Assessments for Hazmat Drivers, January 25.

Congressional Research Service (CRS) (2005b) Tattelman, T.B. Legal Issues Concerning State and Local Authority to Restrict the Transportation of Hazardous Materials by Rail, February 4.

Congressional Research Service (CRS) (2006) Transportation Issues for the 110th Congress.

Congressional Research Service (CRS) (2009a) Transportation Security: Issues for the 111th Congress, January 28.

Congressional Research Service (CRS) (2009b) Transportation Security: Issues for the 111th Congress, May 15.

Congressional Research Service (CRS) (2010) Paul W. Parfomak, Pipeline Safety and Security: Federal Programs, February 18.

Coyle, P. (2009a) Reader Comments, Hazmat routing, Chemical Facility Security News, March 25, http://chemical-facility-security-news.blogspot.com/search?q=hazmat+transport&updated-max=2009-03-25T09:36:00-04:00&max-results=20 (accessed November 18, 2011).

Coyle, P. (2009b) A Closer Look at Hazmat Truck Security, Chemical Facility Security News, April 2, http://chemical-facility-security-news.blogspot.com/2009_04_02_archive.html (accessed November 18, 2011).

Coyle, P. (2010a) Reader Comment, Effectiveness of RTK, Chemical Facility Security News, August 24, http://chemical-facility-security-news.blogspot.com/2010_08_01_archive.html (accessed November 18, 2011).

Coyle, P. (2010b) Ballistic Protection for Hazmat Transport, Chemical Facility Security News, September 24, http://chemical-facility-security-news.blogspot.com/2010_09_01_archive.html (accessed November 18, 2011).

Department of Homeland Security (DHS) (2008) Small Vessel Security Strategy, April.

General Accountability Office (GAO) (2008a) First Responders' Ability to Detect and Model Hazardous Releases in Urban Areas Is Significantly Limited, June.

General Accountability Office (GAO) (2008b) Actions Have Been Taken to Enhance Security, but the Federal Strategy Can Be Strengthened and Security Efforts Better Monitored, April.

General Accountability Office (GAO) (2009a) Federal Efforts to Strengthen Security Should Be Better Coordinated and Targeted.

General Accountability Office (GAO) (2009b) Highway Infrastructure, Federal Efforts to Strengthen Security Should Be Better.

General Accountability Office (GAO) (2010) General Accountability Office (GAO) Highway Infrastructure, Federal Efforts to Strengthen Security Should Be Better Coordinated and Targeted on the Nation's Most Critical Highway Infrastructure, April 2010.

Giermanski, J. (2010) Hazmat by truck: another vulnerability? *Journal of Transportation Security*, **3** (1), 33–39.

Giermanski, J. (2011) What the Rotterdam Rules Should Do for Global Supply Chain Security, http://www.maritime-executive.com/article/what-the-rotterdam-rules-should-do-for-global-supply- chain-security (accessed November 18, 2011).

Harrald, J.R. (2005) Sea trade and security: an assessment of the post 9/11 reaction. *Journal of International Affairs*, **59** (1), 157–179.

Hartong, M., Goel, R., and Wijesekera, D. (2008) Security and the US rail infrastructure. *International Journal of Critical Infrastructure Protection*, **1**, 15–28.

Jenkins, B.M. and Butterworth, R.R. (2010) Potential Terrorist Uses of Highway-Borne Hazardous Materials, Minetra Transportation Institute, January.

Lindquist, E. and Slack, J. (1999) Problems of Hazardous Materials Transport in Texas, TTI, January.

National Conference of State Legislatures (NCSL) (2011–2012) Policies for the Jurisdiction of the Transportation Committee, http://www.ncsl.org/Default.aspx?TabID=773&tabs=854,15,699#699 (accessed November 18, 2011).

National Highway Institute (NHI) and Federal Highway Administration (FHWA) (1996) Highway Routing of Hazardous Materials: Guidelines for Applying Criteria. Arlington, VA, November.

NRC/TRB (2005) Cooperative Research for Hazardous Materials Transportation, TRB special report 283, Washington, DC.

OIG (Office of the Inspector General) (2009) Testimony Number CC-2009-096, PHMSA's Process for Granting Special Permits and Approvals for Transporting Hazardous Materials Raises Safety Concerns, September 10.

OIG (Office of the Inspector General) (2010) Report Number AV-2010-045, New Approaches Needed in Managing PHMSA's Special Permits and Approvals Program, March 4.

Quarterman, C.L. (2010a) Actions Taken and Needed To Improve Management and Oversight of PHMSA's Hazardous Materials Special Permits and Approvals Program. Before the Committee on Transportation and Infrastructure, United States House of Representatives April 22, http://www.phmsa.dot.gov/staticfiles/PHMSA/DownloadableFiles/House%20T&I%20Written%20Statement%20on%20Special%20Permits%20April%2022%202010%20FINAL.pdf (accessed November 18, 2011).

Quarterman, C.L. (2010b) "Remarks" IAFC 2010 International Hazmat Response Teams Conference Baltimore, MD, May 21, http://www.phmsa.dot.gov/staticfiles/PHMSA/DownloadableFiles/Files/Speech%20Files/Quarterman%20-%20IAFC%20Conference%20Remarks%20

%20FINAL%20-%205-17-2010.pdf (accessed November 18, 2011).
Shaver, D.K. and Kaiser, M. (1998) Criteria for highway routing of hazardous materials, TRB, Nat coop highway research program.
TRB (2005) Cooperative Research for Hazardous Materials Transportation.
TSA (2007) Transportation Systems Sector Specific Plan, Annex E, Freight Rail, May.
USDOT (2000) Department-wide Program Evaluation of the Hazardous Materials Transportation Program, March.
USDOT (2008) National Rail Safety Action Plan Final Report 2005–2008.

Web Sites

AAR—http://www.aar.org/
ATA—http://www.truckline.com/Pages/Home.aspx
DHS—http://www.dhs.gov/index.shtm
FMCSA—http://www.fmcsa.dot.gov/
FRA—http://www.fra.dot.gov/
NCSL—http://www.ncsl.org/
PHMSA—http://www.phmsa.dot.gov/
Statemaster—http://www.statemaster.com/index.php
TSA—http://www.tsa.gov/
USDOT—http://www.dot.gov/

15
Security of Hazmat Transports in Iran

Amir Saman Kheirkhah

15.1
Introduction

HAZMATs have high potential to harm human lives; therefore terrorist groups are interested in obtaining these materials in order to use them as weapons of mass destruction. Iran is becoming increasingly exposed to HAZMATs and the related risks: Iran is an oil-producing country and its oil and gas resources must be transported for domestic consumption and export purposes. Meanwhile, Iran is situated on transit route of several Central Asian countries, most of which produce oil. Sabotage in oil transportation systems may have devastating consequences. Such deliberate act of terror is nevertheless very likely to happen, because neighboring countries that provide their fuel through Iran's transportation network are strongly involved in terrorism issues.

Many of the economic and population centers that can be considered as a potential target for terrorists, are origins and/or destinations of hazmat transport vehicles. Refineries, power plants, chemical industries, recycling centers, power-distribution centers and large cities are among these origins and destinations. An intentional incident in these centers can lead to mass casualties. Therefore, such risks must be minimized to the largest possible extent.

Weakness of safety management systems in HAZMAT transportation organizations has resulted in Iranian people being more vulnerable. Some quasi-intentional deadly accidents of HAZMAT transportation have been attributed to weaknesses of safety management systems. A search in the archives of Iran and Hamshahri newspapers showed that in the two months of September and October 1999, 6 accidents took place, the most critical of which led to 20 fatalities and 110 injuries. Incidents such as the sinking of a truck loaded with poisonous materials in a lake behind a water dam, explosion of a fuel tank, and eruption of acid from a truck were reported. The most serious accident took place in 2004. On March 20th of that year, a train loaded with explosives collided with wagons filled with flammable materials at a station in eastern Iran, leading to the death of 283 passengers. In another accident on June 25, 2005 a trailer loaded with fuel hit a number of cars

Security Aspects of Uni- and Multimodal Hazmat Transportation Systems, First Edition. Edited by Genserik L.L. Reniers, Luca Zamparini.

at a checkpoint in southeast Iran leading to 90 deaths and 114 injuries (a detailed overview of these incidents will follow).

A comparative look at the Iran US statistics over one year (i.e. 2004) reveals that the death toll in Iran was several times higher than in the United States, despite the fact that the US's transportation network is several times larger than that of Iran. This difference is not due to a low level of national security. In fact, it clarifies the gap between the safety level of HAZMAT transportation system in Iran and the optimal level.

In this chapter, major events of hazardous-material transportation in Iran are reviewed, and the current safety and security situation is discussed. The strengths and weaknesses of HAZMAT transportation system in Iran are listed, and some strategies for improving the level of safety and security, in this regard, are analyzed.

15.2 Overview of the Current Status

As mentioned in the introduction, the number of accidents involving hazardous materials in Iran remains high relative to the rate of material movement. Fortunately, not many accidents led to the loss of lives or damage to the environment. However, they remained threatening to the lives of many and have even at times led to serious injuries and deaths.

We discern accidents involving hazardous materials based on the intensity of the accidents: low- and high-intensity accidents. It bears mentioning that the information was obtained from newspapers and the lower the impact of the accident, the less exact is the information on it. Due to this, we will focus on high intensity accidents. We will try to analyze these incidents based on existing information on accidents.

The greatest accident recorded in Iran's history of hazardous-materials transportation was the explosion of train wagons carrying dangerous substances in the Khorasan province in 2004. In this incident, 46 wagons carrier of hazardous materials (naphtha, ammonium nitrate, etc . . .) exploded in a railroad station and as a result 283 persons were killed while hundreds were injured. The cause of the accident was the sidetracking of the said wagons at one station and their ultimate explosion in another station. Among factors contributing to the gravity of the accident, the following were stressed: permission issued to enter the station after the fire was initially contained; unawareness of firefighters about the wagon's contents. The use of water for containing the fire had the opposite effect of causing explosions and secondary fires and this was while people had gathered around, thinking that the situation was under control. Another issue was the linking of 46 wagons transporting hazardous materials.

However, one point worthy of mention was the timely information dissemination to passenger trains that followed in averting the accident scene, thus saving 300 lives.

Another major accident took place at a checkpoint in the Sistan-Baluchistan province (in southeast of Iran) in 2005. A speeding gas truck could not come to a halt due to break failure and thus collided severely with buses and cars parked at the checkpoint. The truck caught fire and exploded, as a result of which 90 persons were killed and 114 wounded. In this accident in addition to violations of commitment by the driver (who was speeding in order to reach his destination), the brakes failed to operate and the location of the checkpoint was also inadequate. The intensity of the accident was affected by a number of factors: the inadequate safety technology of gas truck; the sensitive situation of the checkpoint resulting in the long queue of vehicles, and the long-term storage of contraband commodities (such as petrol).

In addition to the Khorasan incident, some other events have been reported in newspapers that have a security nature. In March 2004, two tankers carrying toxic chemicals were overturned near the Qeshlaq dam of Kordestan province. The incident caused 12 000 liters of the toxic material to enter the lake behind the dam. In November 2010, traffic police of Hamadan province seized a tanker with a faked placard. This tanker contained acidic hazardous materials and had unlawfully entered an urban area. In December 2010, 11 trucks containing flammable liquid (petrol) burned in a fire on the Iran–Iraq border.

Development and implementation of safety management systems, at the national and at the organizational level, can prevent the occurrence of such incidents, and can help the security forces to react timely to terroristic activities.

In the case of three accidents involving 50 injuries each, one incident involved the explosion of a liquid-gas bunker and the two other cases were caused by nonstandard trailers carrying corrosive acids. In all three cases, reckless driving was the primary cause of the accident and the intensity of the gas-bunker accident was due to the fact that it took place in a heavy traffic zone. In the case of the trailers, inadequate and nonstandard packaging and loading of materials caused the accidents. In analyzing the factors in the above case and other accidents it bears mentioning that the separation of factors, that is, driver, network, vehicle, loading, and other variables does not denote that these factors are independent from one another, as inadequate networks increase the frequency of driving violations or inadequate planning can lead to greater use of dangerous routes on the network. In addition, in the transportation of hazardous materials, dangers in transportation planning were overlooked.

Unsuitable planning does not cause accidents by itself but can increase the chance of accidents taking place. For instance, in the Sistan-Baluchistan accident, the inappropriate transportation schedule led to a driving violation and hence to the accident. In the case of the truck that fell behind the dam, the question arises as to why such a transportation route was selected. A question that can be raised at this point is whether the adoption of preventive strategies can prevent the recurrence of past accidents. Can these strategies help contain the negative consequences of hazardous materials' transportation? To answer the above questions we will show that the long-term effects of adopting these strategies can help prevent the recurrence of such problems and will therefore reduce the frequency and intensity of accidents associated with the transportation of hazardous materials. The continuous improvement strategy would be to improve the standards of the driver's skills, loading methods, and the transportation operations. By improving these standards and their enforcement the frequency of the following factors will decline: accidents caused by driver's mistakes, leaking of dangerous substances in the aftermath of accidents, occurrence of accidents in highly populated areas or high-traffic routes, accidents caused by using dangerous roads.

As pointed out previously, the main causes of accidents are driver's mistakes and use of nonstandard holders and tankers for hazardous materials. For instance, in the case of the Sistan-Baluchistan accident the main contributors were the driver's speed, low safety standards of the fuel tank, and bad scheduling of transportation and selection of inadequate routes.

By adopting the foregoing strategy, the probability of accident occurrence will decrease. With regard to the foregoing example, one of the key factors was the necessity of petrol transfer by tanker, which could have been carried out by pipelines if the pertinent strategy had been selected (gas transfer, following a strategy of encouraging usage of gas combustion engines and thus reducing the need for the transfer of petrol and thereby containing the contraband of this commodity). The adoption of this strategy will reduce the need for the transportation of liquid gas by bunkered trucks, limiting the probability of accidents and terrorist actions for this type of vehicles, especially the kind of accidents similar to the one that led to 20 deaths.

One of the main causes of accidents in the past was lack of attention and awareness about the dangers associated with the transportation of hazardous materials. Had people been aware of the dangers of hazardous-materials transportation, they would have exerted more care when approaching or passing vehicles carrying hazardous materials. In such a case the frequency of these accidents would have been considerably reduced. Or in case of the Neishabour accident, instead of gathering around the location of the accident, they would have cooperated with fire fighters; or else, they could have prevented the movement of dangerous materials in the vicinity of their houses, thus forcing local officials to observe safety measures.

Lack of attention to adequate safety measures is due to transportation companies not using safety-management systems. Had such systems been used, each activity would have been measured and commensurate resource would have been allocated to it according to adequate planning. In the case of the Neishabour accident, nonobservance of safety measures regarding the transportation of hazardous materials and inattention of railway guards led to the sidetracking of the wagons and the ensuing accident. Also, the timely notifications about the impeding danger led to the rescue of hundreds of lives and this very factor underscores the sensitivity of using safety systems leading to adequate and timely policies.

Another role of the safety systems is the gathering of and access to accurate transportation information. If firefighters had adequate knowledge of the hazardous substance they could have taken more effective measures in containing the blaze and secondary explosions of the wagons could have been effectively prevented.

The probability of adequate management and implementation of the foregoing strategies can only increase if specialized transportation firms undertake the movement of hazardous materials. As such, not only can better controls be implemented, but the process can become more economical. This would be possible only through the adoption of a suitable strategy.

In the final analysis, it is obvious that better and more effective strategies are only possible on the basis of better knowledge and information. This very fact in itself supports the necessity of research and study in the field of transportation of hazardous materials.

15.3
Strengths and Weaknesses of Iran's Transportation System

Based on the above accidents' review and discussion, a list of strengths and weaknesses of Iran's transportation systems for transporting hazardous materials is presented in this section, based on which suitable strategies can be developed.

- Strengths:
  - S1: Devising new regulations and government's increasing focus on the transportation of hazardous materials over recent years.
  - S2: Separation of government (monitoring role) and the private sector (contractor) in the transportation field as opposed to other economic sectors in Iran that are governmental.
  - S3: Implementation of a number of training programs for road-transportation companies over the past few years.
  - S4: Introduction and implementation of new technologies (GPS, ITS, GIS . . .).
  - S5: Greater public opinion sensitivity towards environmental issues.
  - S6: Increased level of technology and standards adopted by auto manufacturers in Iran.
  - S7: Vast plans of railway development throughout Iran.
  - S8: Government's established duty in resolving accident-prone areas of transportation network.
- Weaknesses:
  - W1: Numerous deadly accidents over past years (Iran's roads are highly unsafe).
  - W2: Obsolescence of transportation fleet.

- W3: Absence of or weak safety-management systems in companies transporting dangerous materials.
- W4: Weakness of detailed operational plans (such as danger-measurement methods and danger-level criteria definition).
- W5: Weak research organization and information gathering on transportation of hazardous materials.
- W6: Due to uncontrolled increase in number of automobiles, Iran's road network capacity is not enough for establishing secure transportations.
- W7: Slowly developing safety systems and law, versus sharp development of industrial sectors, and especially the petrochemical and chemical industries.
- W8: Private ownerships of trucks reducing the chance of legal follow up further to accidents (involving the death of the driver/owner).

15.4 Safety and Security Strategies

Along with analyzing weaknesses (and also strengths) of safety and security of transportation systems in Iran, in Section 15.3, we point to some strategies that can underpin governmental efforts to reduce hazardous-materials transportation risks. In this section, a more complete list of safety and security development strategies is provided. For each strategy, the responsible organization (or organizations) is (are) also introduced. Table 15.1 contains these strategies.

- *Strategy 1:* Continuous improvement of drivers' skills standards, truck specs, loading methods, packaging, and the transportation operations. Responsible organizations: Road and Transportation Ministry/Institute of Standard and

Table 15.1 Strategies and related strengths and weaknesses.

Row	Strategy	Strengths								Weaknesses							
		S1	S2	S3	S4	S5	S6	S7	S8	W1	W2	W3	W4	W5	W6	W7	W8
1	Strategy 1	*		*			*					*	*				
2	Strategy 2									*	*				*	*	
3	Strategy 3	*			*		*								*		
4	Strategy 4					*				*							
5	Strategy 5	*								*		*					
6	Strategy 6		*									*					*
7	Strategy 7							*	*				*	*			

*, related strengths and weaknesses.

Industrial Research of Iran (The Institute of Standards & Industrial Research of Iran, is the sole organization in the country that can lawfully develop and designate official standards for products. It is also the responsible body for conducting them through the endorsement of the Council of Compulsory Standards).

- *Strategy 2:* construction of pipelines for the transportation of oil derivatives. Responsible organization: National Iranian Oil Company (National Iranian Oil Company [NIOC]) is directing and making policies for exploration, drilling, production, research and development, refining, distribution and export of oil, gas, petroleum products).
- *Strategy 3:* Necessity of using new standards and technologies in hazardous-materials transportation. Responsible organization: Iran Road Maintenance and Transportation Organization (RMTO). Duties of State Transportation and Terminals Organization are:
  - Regulating guidelines, policies and programs of road transportation and coordinating various sectors involved in transportation as well as monitoring and controlling road transportation affairs and providing necessary proposals in order to observe economic productivity and facilitate communications at the time of ratifying statewide road construction projects.
  - Constructing, utilizing, expanding and maintaining complexes for goods terminals and other required installations such as TIR parking lots and other *en-route* general and welfare service complexes as well as centers for traffic distribution, control and monitor.
  - Monitoring passenger-transportation companies in terminals to ensure correct implementation of passenger-transportation rules and regulations.
  - Provision and regulation of bills, ratifications, bylaws and necessary directives regarding local and international transportation affairs and presenting them to State Supreme Transportation Coordination Council and other authorities to confirm, ratify and pursue ratification.
  - Issuing activity permits to road transportation companies and other necessary permits and authorizations in local and international transportation activities.
  - Provision of necessary arrangements to facilitate transit and export affairs.
  - Provision of necessary accommodations to facilitate promotion of safety level and services as well as decreasing environmental damages in statewide road-transportation sector.
  - Review and determination of type, number and specifications of existing and required transportation fleet for the road-transportation sector and also cooperation in provision and distribution of such a fleet.

- Centralization of fundamental road transportation data and statistics in cooperation with related organizations and institutions and establishment of data bank.
- Issuing permit for construction of *en-route* welfare service complexes and installations by public, cooperative and private sector.
- Issuing permit for special utilization of statewide roadway network such as transit permits to foreign vehicles and special permits for traffic loads in accordance to relevant rules and regulations.
- Scheduling special and general training programs for employees of the organization and providing necessary ground for developing required specialties for a statewide road-transportation fleet through local and foreign training centers.
- Issuing relevant permissions and official approvals for establishment and utilizing technical inspection centers of public rural light and heavy goods/passengers vehicles.
- Carrying out commercial affairs on the road-transportation sector towards fundamental duties and objectives in the framework of relevant rules and regulations.

• *Strategy 4:* Media coverage of dangers associated with the transportation of hazardous materials. Responsible organization: Ministry of Road and Transportation.

• *Strategy 5:* Necessity of establishing and using safety management systems and security management program in hazardous-materials transportation companies. Responsible organization: RMTO.

• *Strategy 6:* Creation and organization of specialized hazardous-materials transportation companies instead of hiring regular transportation companies and singular trucks. Responsible organizations: state bank and RMTO.

• *Strategy 7:* Creation of hazardous materials research center and using research findings in the country's road construction and hazardous-materials transportation planning. Responsible organizations: Ministry of Sciences, Researches and Technology (MSRT), and Ministry of Road and Transportation.

15.5 Discussion

Two of the main problems are lack of motivation and lack of training among truck drivers and other personnel associated with hazardous-materials transportation. In addition to methods and systems that are not standardized, there are limited standardized regulations in this area that have been mainly adapted and translated.

Inspired by the theory of continuous improvement, this strategy suggests that the level of these standards must be properly set and developed. Fortunately, over the past years a number of training courses was organized in this area and the country's technological level has improved and the government is responsible by law to follow up on these activities in order to contain danger on roads. Terrorists often exploit the weaknesses of safety and security systems. By using strategy 1, the possibilities open to terrorists will be more and more reduced.

Over the past few years, the number of automobiles has increased considerably and the heavy traffic has spread to intercity highways and roads. In addition, in view of industrial development, freight movement has been on the rise. Increased consumption of energy further exacerbates this problem. Oil and commodity transit to and from Central Asia are expanding. In light of this, the construction of oil pipelines for the transfer of oil would be economically advisable and can contribute to risk reduction on the roads. In addition, the growing number of automobiles necessitates increased gas imports, requiring the movement of gas/petrol tankers on the road and thereby enhancing risk.

Fortunately, introduction of new communication technologies such as GIS, GPS and ITS over the past years and the utilization of novel ABS brake technology have led to reduced human error in prompting accidents. In addition, observation of new environmental standards and quality by the auto industry together with road-construction efforts lead to risk reduction. The government must enforce the use of these standards and technologies throughout the transportation system, especially when it comes to the transportation of hazardous materials in order to decrease risks. GPS allows police to track the hazardous materials carriers and respond in time to terrorist actions.

Creating public awareness is a suitable means for furthering policies. Over the past years, Iranian media and papers have been able to create an awareness wave and thus force policymakers to implement their suggested political and/or social program. Given the high mortality rate associated with hazardous-materials transportation in Iran, public media can be encouraged to create public awareness on this issue so that the government and parliament can allocate sufficient budget for the reduction of risk associated with the transportation of hazardous materials.

The fact that transportation firms forego the usage of safety-management systems is one of the main risk factors. Despite new regulations, the government has not paid sufficient attention to this issue that deserves due emphasis.

One of the main problems besetting Iran's transportation system is that trucks are used by their owners and since in 50% of the accidents the driver is killed, there is no way to make legal follow ups. It is in this light that the government seeks to organize these trucks in the context of firms. Given the specific attributes of hazardous materials it would be best for qualified firms to handle safety-management standards and systems.

The issue of reducing risk associated with hazardous-materials transportation has received due attention at the policy level, while it remains under researched in Iran. Therefore, it is necessary for the Ministry of Science, Research and

Technology to create a research center in this area and incorporate research findings into the decision-making process affecting long-term investment in road and railway construction.

15.6 Conclusions

The issues of hazardous-materials transport security and safety in Iran were addressed in this chapter. Being an oil-rich and developing country, and Iran's geographical position in the Middle East, have led to Iranian people being seriously at risk.

Iran's current situation was overviewed, it was mentioned that the number of accidents involving hazardous materials in Iran was high relative to the rate of material movement. Some of these incidents had security aspects.

After analysis of the incidents, the strengths and weaknesses of Iran's transportation-related systems were listed and strategies for improving the present situation were introduced.

The suggested strategies will have to be implemented and evaluated in order to be further improved in the future.

Certainly, in addition to strategic plans, tactical plans and short-term programs also have key roles in reducing the risk. Therefore, the following areas are suggested for future research:

- Development of decision support systems for routing and scheduling hazardous-materials transportation on Iran's road network.
- Optimal location finding for facilities generating hazardous-materials transportations.
- Using operational-research techniques in solving security-related problems.

Bibliography

Colen, M.S. (1997) Regulation of hazardous waste transportation: Federal, state, local, or all of the above. *Journal of Hazardous Materials*, **15** (1–2), 7–56.

Ghazinoory, S. and Huisingh, D. (2006) National program for cleaner production (CP) in Iran: a framework and draft. *Journal of Cleaner Production*, **14**, 194–200.

Ghazinoory, S. and Kheirkhah, A.S. (2008) Transportation of hazardous materials in Iran: a strategic approach for decreasing accidents. *Transport*, **23** (2), 104–111.

Kheirkhah, A.S., Esmailzadeh, A., and Ghazinoory, S. (2009) Developing strategies to reduce the risk of hazardous materials transportation in Iran using the method of fuzzy SWOT analysis. *Transport*, **24** (4), 325–332.

16
Conclusions and Recommendations

Genserik Reniers and Luca Zamparini

The topics and the structure of this book have evolved from a twofold consideration. It appears that the large majority of the state-of-the-art research related to security is currently carried out on single transportation modes (regarding, for example, routing and rerouting problems, GIS-based programs, conceptualizations, different kinds of analyses, real-time information systems). Moreover, the developed and/or proposed academic methodologies are hardly used in industrial practice to tackle transport security optimization. Translating complex transportation problems with respect to security into understandable, usable, and very concrete managerial approaches is a true challenge for the academic world. However, it is also a challenge for the industrial transport sector, since practitioners need to fully recognize their sector is subject to substantial threats. At present, disentangled security measures are taken in industry (often legislation-driven), but a clear proactive vision is lacking and long-term voluntary engagements are largely missing. There is no structural approach to continuously improve security from a systems viewpoint. The industry's recognition should therefore transcend words, and advanced managerial actions should be taken from a systems perspective. Security as a management domain needs to be advanced by the industrial world and security investments need to be taken wherever and whenever required. To this end, more intensive and long-term collaborative arrangements should be elaborated and fostered between researchers and practitioners.

This book clearly demonstrates the need for more academic research on the multimodality aspect of hazmat transports' security. The existence of different transportation modes obviously increases the freedom of choice for shippers. Hence, an opportunity is readily available to investigate a number of essential parameters for determining several options to ship different amounts of various hazardous goods. Next to parameters such as cost and safety, security should be one of the parameters to base the prioritization of the choice of the shipment alternative. However, in industrial practice, shippers hardly ever use "security" as an important parameter to investigate their options for organizing hazmat transports. One of the reasons for this fact is the lack of available methods for making multimodal transportation choices: thus far the topic is simply underresearched.

Security Aspects of Uni- and Multimodal Hazmat Transportation Systems, First Edition. Edited by Genserik L.L. Reniers, Luca Zamparini.

Although multimodal transportation is traditionally investigated from a cost- or safety-based point of view by academics, the fundamental research field of security is slowly gaining importance worldwide. Nonetheless, there is still a lot of progress to be made. For example, Bayesian network approaches need to be developed, game-theoretical studies should be carried out, metaheuristics should be elaborated, multicriteria modeling should be employed, etc. Based on such academic methodologies for advancing multimodal security, managerial models may be developed, in turn leading to actions taken by practitioners to create efficient, cost-effective, safe and secure multimodal hazmat transport supply chains worldwide.

16.1 Unimodal and Multimodal Transportation Put into Perspective

As previously stated, industrial practice is mainly aimed at protecting single transport modes, without looking at the systemic and multimodal picture of hazmat transportation. It is evident that single-mode transport security is indeed very important to avoid security incidents as much as possible as well as to contain and/or decrease possible consequences. Nonetheless, looking at hazmat transportation security from a multimodal perspective brings novel – more efficient – risk-management approaches to the front, which are indispensible to create effective hazmat-transportation security.

Since road transports count for the lion's share of hazmat shipments, security awareness and security efforts in this sector are predominantly present. Routing solutions and advanced technologies seem to be the main drivers for enhancing security in this transportation mode in present industrial settings.

Railway transportation thus far has been very safe and secure. Nevertheless, since 9/11, intentional risks have become important to consider in the railroad industry. Hazmat railroad tankers might be used as weapons, or they might form a target causing mass casualties. The fact that plans were apparently being made by Al Qaeda to attack the US railroad system (according to documents found in Osama Bin Laden's hiding location), indicates that terrorist attack scenarios are indeed very real for this transport mode.

The inland waterway transportation mode is known to be very secure, and only few hazmat security incidents are known. Nevertheless, since a well-organized attack on this transport mode could have a significant economic impact, as well as lead to a large number of fatalities, technological and managerial security measures should be further advanced.

Pipeline security differs between different geographic regions and different socioeconomic environments. Therefore, security measures should be based on country-by-country considerations, and security resources should be assigned based on local information and effective international coordination.

Making hazmat transportation more secure through multimodal solutions is not an easy task. Studies regarding multimodal transport optimization and hazmat threat-assessment effectiveness are still at an early stage. Looking at existing multimodal transport hubs, it is obvious that, due to the possible extremely severe consequences of incidents, they should be defended against potential malicious acts, with focus on radioactive materials, toxic products, as well as all other hazardous materials being handled in such locations.

16.2 A Country-Wise Comparative Study

The contributions originating from countries belonging to different continents around the world, clearly indicate that no standardized or harmonized practices are currently applied in hazmat-transport security.

In Europe, the cases of Italy and the Netherlands were discussed. In Italy, securing hazmat transports is regarded as a serious matter. Several security domains regarding such transportation need to be dealt with, for example, toxic waste dumping, transportation of explosives, radioactive materials transports, and cargo theft. It should also be noticed that many of the hazmat-transport security events that took place in the last years were related to international transport activities aiming to smuggle or illegally trade these substances.

In The Netherlands, security issues related to hazmat transports are mainly limited to cargo theft. Terrorist actions are extremely rare. Although focusing on the prevention of cargo theft, the country has also taken a variety of measures to prevent a terrorist-induced disaster and to handle terrorist attacks in the eventuality that this would be needed.

In the United States, hazmat transportation security is considered as a domain of great concern. Security gaps are traced and dealt with. Efforts are continuously being made to lower the risk of mass casualties or large financial losses resulting from a terrorist attack, thereby taking costs and effectiveness of security measures into consideration.

Due to several parameters such as, for example, the geographical location, transporting hazardous materials represents a considerable security risk in Iran. Many weaknesses in the transportation systems still exist and practices and strategies are being suggested and developed to tackle these potential threats.

Although a differentiation of security measures between different countries is clearly needed to some extent, some general conclusions and recommendations can be derived. The next section discusses these region-independent possible ways of improvement.

16.3 A Look into the Future: Sustainable Multimodality

A strive for continuous improvements is essential to the continued safety and security successes of hazmat transportation. Progress should be based on consistent and pro-active efforts, rather than on *ad hoc* and reactive changes. Without continuously improving efforts, the industrial hazmat transportation sector might become complacent, eventually leading to lower safety and security standards. Although the current safety and security record is the direct result of previous and current investments, continuous improvements – and the accompanying investments – should be aimed for at all times. Hazmat transporters need to monitor and track emerging security trends so that they can be proactive, preventive, and/or respond whenever necessary. As population figures continue to grow throughout the world, and the distance between the public and the hazardous-material transportation operations diminish, society's expectations of what is an acceptable risk may change. This in turn may impact the existing handling and transportation approaches of hazardous materials, immediately affecting the companies' daily operations.

The future of hazmat transportation seems to be heading towards multimodality and more collaboration, taking vital parameters such as cost effectiveness, environmental aspects, and safety and security into account. *Sustainable multimodality* seems to be the answer to the problems of today. To attain this goal, many research and development opportunities exist today for academics and plenty of financial benefits may be grasped by practitioners.

Three important policy implications may be put forward to ultimately attain the ambitious objective of sustainable multimodality. First, policy measures need to be taken to stimulate collaborative arrangements concerning hazmat transportation between academia and industry, in order to enhance knowledge-based multimodal solutions and in order to further optimize existing hazmat-transport security measures. Secondly, hazmat logistics service providers and shippers of hazardous materials should be urged to cooperate and to make multimodal agreements whenever increasing supply chain efficiencies, and best-available practices should be exchanged more intensively and more efficiently. Thirdly, hazmat-transport security risk policies within different countries located within similar world regions should be harmonized and standardized for all transportation modes, and while elaborating a single transport mode's security standard, the other modes' security standards should be taken into account.

Index

Security Aspects of Uni- and Multimodal Hazmat Transportation Systems, First Edition. Edited by Genserik L.L. Reniers, Luca Zamparini.